United States
Environmental Protection Agency

EPA/600/R-00/057
September 2000

EPANET 2

USERS MANUAL

By

Lewis A. Rossman
Water Supply and Water Resources Division
National Risk Management Research Laboratory
Cincinnati, OH 45268

NATIONAL RISK MANAGEMENT RESEARCH LABORATORY
OFFICE OF RESEARCH AND DEVELOPMENT
U.S. ENVIRONMENTAL PROTECTION AGENCY
CINCINNATI, OH 45268

DISCLAIMER

The information in this document has been funded wholly or in part by the U.S. Environmental Protection Agency (EPA). It has been subjected to the Agency's peer and administrative review, and has been approved for publication as an EPA document. Mention of trade names or commercial products does not constitute endorsement or recommendation for use.

Although a reasonable effort has been made to assure that the results obtained are correct, the computer programs described in this manual are experimental. Therefore the author and the U.S. Environmental Protection Agency are not responsible and assume no liability whatsoever for any results or any use made of the results obtained from these programs, nor for any damages or litigation that result from the use of these programs for any purpose.

FOREWORD

The U.S. Environmental Protection Agency is charged by Congress with protecting the Nation's land, air, and water resources. Under a mandate of national environmental laws, the Agency strives to formulate and implement actions leading to a compatible balance between human activities and the ability of natural systems to support and nurture life. To meet this mandate, EPA's research program is providing data and technical support for solving environmental problems today and building a science knowledge base necessary to manage our ecological resources wisely, understand how pollutants affect our health, and prevent or reduce environmental risks in the future.

The National Risk Management Research Laboratory is the Agency's center for investigation of technological and management approaches for reducing risks from threats to human health and the environment. The focus of the Laboratory's research program is on methods for the prevention and control of pollution to the air, land, water, and subsurface resources; protection of water quality in public water systems; remediation of contaminated sites and ground water; and prevention and control of indoor air pollution. The goal of this research effort is to catalyze development and implementation of innovative, cost-effective environmental technologies; develop scientific and engineering information needed by EPA to support regulatory and policy decisions; and provide technical support and information transfer to ensure effective implementation of environmental regulations and strategies.

In order to meet regulatory requirements and customer expectations, water utilities are feeling a growing need to understand better the movement and transformations undergone by treated water introduced into their distribution systems. EPANET is a computerized simulation model that helps meet this goal. It predicts the dynamic hydraulic and water quality behavior within a drinking water distribution system operating over an extended period of time. This manual describes the operation of a newly revised version of the program that has incorporated many modeling enhancements made over the past several years.

E. Timothy Oppelt, Director
National Risk Management Research Laboratory

(This page intentionally left blank.)

CONTENTS

CHAPTER 1 - INTRODUCTION 9
- 1.1 WHAT IS EPANET 9
- 1.2 HYDRAULIC MODELING CAPABILITIES 9
- 1.3 WATER QUALITY MODELING CAPABILITIES 10
- 1.4 STEPS IN USING EPANET 11
- 1.5 ABOUT THIS MANUAL 11

CHAPTER 2 - QUICK START TUTORIAL 13
- 2.1 INSTALLING EPANET 13
- 2.2 EXAMPLE NETWORK 13
- 2.3 PROJECT SETUP 15
- 2.4 DRAWING THE NETWORK 16
- 2.5 SETTING OBJECT PROPERTIES 18
- 2.6 SAVING AND OPENING PROJECTS 20
- 2.7 RUNNING A SINGLE PERIOD ANALYSIS 20
- 2.8 RUNNING AN EXTENDED PERIOD ANALYSIS 21
- 2.9 RUNNING A WATER QUALITY ANALYSIS 24

CHAPTER 3 - THE NETWORK MODEL 27
- 3.1 PHYSICAL COMPONENTS 27
- 3.2 NON-PHYSICAL COMPONENTS 34
- 3.3 HYDRAULIC SIMULATION MODEL 40
- 3.4 WATER QUALITY SIMULATION MODEL 41

CHAPTER 4 - EPANET'S WORKSPACE 47
- 4.1 OVERVIEW 47
- 4.2 MENU BAR 48
- 4.3 TOOLBARS 51
- 4.4 STATUS BAR 52
- 4.5 NETWORK MAP 53
- 4.6 DATA BROWSER 53
- 4.7 MAP BROWSER 54
- 4.8 PROPERTY EDITOR 54
- 4.9 PROGRAM PREFERENCES 55

CHAPTER 5 - WORKING WITH PROJECTS 59
- 5.1 OPENING AND SAVING PROJECT FILES 59
- 5.2 PROJECT DEFAULTS 60
- 5.3 CALIBRATION DATA 62
- 5.4 PROJECT SUMMARY 64

CHAPTER 6 - WORKING WITH OBJECTS 65

    6.1 Types of Objects 65
    6.2 Adding Objects 65
    6.3 Selecting Objects 67
    6.4 Editing Visual Objects 67
    6.5 Editing Non-Visual Objects 74
    6.6 Copying and Pasting Objects 79
    6.7 Shaping and Reversing Links 80
    6.8 Deleting an Object 81
    6.9 Moving an Object 81
    6.10 Selecting a Group of Objects 81
    6.11 Editing a Group of Objects 82

CHAPTER 7 - WORKING WITH THE MAP 83

    7.1 Selecting a Map View 83
    7.2 Setting the Map's Dimensions 84
    7.3 Utilizing a Backdrop Map 85
    7.4 Zooming the Map 86
    7.5 Panning the Map 86
    7.6 Finding an Object 87
    7.7 Map Legends 87
    7.8 Overview Map 89
    7.9 Map Display Options 89

CHAPTER 8 - ANALYZING A NETWORK 93

    8.1 Setting Analysis Options 93
    8.2 Running an Analysis 98
    8.3 Troubleshooting Results 98

CHAPTER 9 - VIEWING RESULTS 101

    9.1 Viewing Results on the Map 101
    9.2 Viewing Results with a Graph 103
    9.3 Viewing Results with a Table 112
    9.4 Viewing Special Reports 115

CHAPTER 10 - PRINTING AND COPYING 121

    10.1 Selecting a Printer 121
    10.2 Setting the Page Format 121
    10.3 Print Preview 122
    10.4 Printing the Current View 122
    10.5 Copying to the Clipboard or to a File 123

CHAPTER 11 - IMPORTING AND EXPORTING 125

    11.1 Project Scenarios 125
    11.2 Exporting a Scenario 125
    11.3 Importing a Scenario 126
    11.4 Importing a Partial Network 126
    11.5 Importing a Network Map 127
    11.6 Exporting the Network Map 127
    11.7 Exporting to a Text File 128

CHAPTER 12 - FREQUENTLY ASKED QUESTIONS 131

APPENDIX A - UNITS OF MEASUREMENT 135

APPENDIX B - ERROR MESSAGES 137

APPENDIX C - COMMAND LINE EPANET 139
    C.1   GENERAL INSTRUCTIONS 139
    C.2   INPUT FILE FORMAT 139
    C.3   REPORT FILE FORMAT 178
    C.4   BINARY OUTPUT FILE FORMAT 181

APPENDIX D - ANALYSIS ALGORITHMS 187
    D.1   HYDRAULICS 187
    D.2   WATER QUALITY 193
    D.3   REFERENCES 199

(This page intentionally left blank.)

CHAPTER 1 - INTRODUCTION

1.1 What is EPANET

EPANET is a computer program that performs extended period simulation of hydraulic and water quality behavior within pressurized pipe networks. A network consists of pipes, nodes (pipe junctions), pumps, valves and storage tanks or reservoirs. EPANET tracks the flow of water in each pipe, the pressure at each node, the height of water in each tank, and the concentration of a chemical species throughout the network during a simulation period comprised of multiple time steps. In addition to chemical species, water age and source tracing can also be simulated.

EPANET is designed to be a research tool for improving our understanding of the movement and fate of drinking water constituents within distribution systems. It can be used for many different kinds of applications in distribution systems analysis. Sampling program design, hydraulic model calibration, chlorine residual analysis, and consumer exposure assessment are some examples. EPANET can help assess alternative management strategies for improving water quality throughout a system. These can include:

- altering source utilization within multiple source systems,
- altering pumping and tank filling/emptying schedules,
- use of satellite treatment, such as re-chlorination at storage tanks,
- targeted pipe cleaning and replacement.

Running under Windows, EPANET provides an integrated environment for editing network input data, running hydraulic and water quality simulations, and viewing the results in a variety of formats. These include color-coded network maps, data tables, time series graphs, and contour plots.

1.2 Hydraulic Modeling Capabilities

Full-featured and accurate hydraulic modeling is a prerequisite for doing effective water quality modeling. EPANET contains a state-of-the-art hydraulic analysis engine that includes the following capabilities:

- places no limit on the size of the network that can be analyzed
- computes friction headloss using the Hazen-Williams, Darcy-Weisbach, or Chezy-Manning formulas
- includes minor head losses for bends, fittings, etc.
- models constant or variable speed pumps
- computes pumping energy and cost

- models various types of valves including shutoff, check, pressure regulating, and flow control valves
- allows storage tanks to have any shape (i.e., diameter can vary with height)
- considers multiple demand categories at nodes, each with its own pattern of time variation
- models pressure-dependent flow issuing from emitters (sprinkler heads)
- can base system operation on both simple tank level or timer controls and on complex rule-based controls.

1.3 Water Quality Modeling Capabilities

In addition to hydraulic modeling, EPANET provides the following water quality modeling capabilities:

- models the movement of a non-reactive tracer material through the network over time
- models the movement and fate of a reactive material as it grows (e.g., a disinfection by-product) or decays (e.g., chlorine residual) with time
- models the age of water throughout a network
- tracks the percent of flow from a given node reaching all other nodes over time
- models reactions both in the bulk flow and at the pipe wall
- uses n-th order kinetics to model reactions in the bulk flow
- uses zero or first order kinetics to model reactions at the pipe wall
- accounts for mass transfer limitations when modeling pipe wall reactions
- allows growth or decay reactions to proceed up to a limiting concentration
- employs global reaction rate coefficients that can be modified on a pipe-by-pipe basis
- allows wall reaction rate coefficients to be correlated to pipe roughness
- allows for time-varying concentration or mass inputs at any location in the network
- models storage tanks as being either complete mix, plug flow, or two-compartment reactors.

By employing these features, EPANET can study such water quality phenomena as:
- blending water from different sources
- age of water throughout a system
- loss of chlorine residuals
- growth of disinfection by-products
- tracking contaminant propagation events.

1.4 Steps in Using EPANET

One typically carries out the following steps when using EPANET to model a water distribution system:

1. Draw a network representation of your distribution system (see Section 6.1) or import a basic description of the network placed in a text file (see Section 11.4).
2. Edit the properties of the objects that make up the system (see Section 6.4)
3. Describe how the system is operated (see Section 6.5)
4. Select a set of analysis options (see Section 8.1)
5. Run a hydraulic/water quality analysis (see Section 8.2)
6. View the results of the analysis (see Chapter 9).

1.5 About This Manual

Chapter 2 of this manual describes how to install EPANET and offers up a quick tutorial on its use. Readers unfamiliar with the basics of modeling distribution systems might wish to review Chapter 3 first before working through the tutorial.

Chapter 3 provides background material on how EPANET models a water distribution system. It discusses the behavior of the physical components that comprise a distribution system as well as how additional modeling information, such as time variations and operational control, are handled. It also provides an overview of how the numerical simulation of system hydraulics and water quality performance is carried out.

Chapter 4 shows how the EPANET workspace is organized. It describes the functions of the various menu options and toolbar buttons, and how the three main windows – the Network Map, the Browser, and the Property Editor—are used.

Chapter 5 discusses the project files that store all of the information contained in an EPANET model of a distribution system. It shows how to create, open, and save these files as well as how to set default project options. It also discusses how to register calibration data that are used to compare simulation results against actual measurements.

Chapter 6 describes how one goes about building a network model of a distribution system with EPANET. It shows how to create the various physical objects (pipes, pumps, valves, junctions, tanks, etc.) that make up a system, how to edit the properties of these objects, and how to describe the way that system demands and operation change over time.

Chapter 7 explains how to use the network map that provides a graphical view of the system being modeled. It shows how to view different design and computed parameters in color-coded fashion on the map, how to re-scale, zoom, and pan the map, how to locate objects on the map, and what options are available to customize the appearance of the map.

Chapter 8 shows how to run a hydraulic/water quality analysis of a network model. It describes the various options that control how the analysis is made and offers some troubleshooting tips to use when examining simulation results.

Chapter 9 discusses the various ways in which the results of an analysis can be viewed. These include different views of the network map, various kinds of graphs and tables, and several different types of special reports.

Chapter 10 explains how to print and copy the views discussed in Chapter 9.

Chapter 11 describes how EPANET can import and export project scenarios. A scenario is a subset of the data that characterizes the current conditions under which a pipe network is being analyzed (e.g., consumer demands, operating rules, water quality reaction coefficients, etc.). It also discusses how to save a project's entire database to a readable text file and how to export the network map to a variety of formats.

Chapter 12 answers questions about how EPANET can be used to model special kinds of situations, such as modeling pneumatic tanks, finding the maximum flow available at a specific pressure, and modeling the growth of disinfection by-products.

The manual also contains several appendixes. Appendix A provides a table of units of expression for all design and computed parameters. Appendix B is a list of error message codes and their meanings that the program can generate. Appendix C describes how EPANET can be run from a command line prompt within a DOS window, and discusses the format of the files that are used with this mode of operation. Appendix D provides details of the procedures and formulas used by EPANET in its hydraulic and water quality analysis algorithms.

CHAPTER 2 - QUICK START TUTORIAL

This chapter provides a tutorial on how to use EPANET. If you are not familiar with the components that comprise a water distribution system and how these are represented in pipe network models you might want to review the first two sections of Chapter 3 first.

2.1 Installing EPANET

EPANET Version 2 is designed to run under the Windows 95/98/NT operating system of an IBM/Intel-compatible personal computer. It is distributed as a single file, **en2setup.exe**, which contains a self-extracting setup program. To install EPANET:

1. Select **Run** from the Windows Start menu.
2. Enter the full path and name of the **en2setup.exe** file or click the **Browse** button to locate it on your computer.
3. Click the **OK** button type to begin the setup process.

The setup program will ask you to choose a folder (directory) where the EPANET files will be placed. The default folder is **c:\Program Files\EPANET2**. After the files are installed your Start Menu will have a new item named EPANET 2.0. To launch EPANET simply select this item off of the Start Menu, then select EPANET 2.0 from the submenu that appears. (The name of the executable file that runs EPANET under Windows is **epanet2w.exe**.)

Should you wish to remove EPANET from your computer, you can use the following procedure:

1. Select **Settings** from the Windows Start menu.
2. Select **Control Panel** from the Settings menu.
3. Double-click on the **Add/Remove Programs** item.
4. Select EPANET 2.0 from the list of programs that appears.
5. Click the **Add/Remove** button.

2.2 Example Network

In this tutorial we will analyze the simple distribution network shown in Figure 2.1 below. It consists of a source reservoir (e.g., a treatment plant clearwell) from which water is pumped into a two-loop pipe network. There is also a pipe leading to a storage tank that floats on the system. The ID labels for the various components are shown in the figure. The nodes in the network have the characteristics shown in Table 2.1. Pipe properties are listed in Table 2.2. In addition, the pump (Link 9) can deliver 150 ft of head at a flow of 600 gpm, and the tank (Node 8) has a 60-ft diameter, a 3.5-ft water level, and a maximum level of 20 feet.

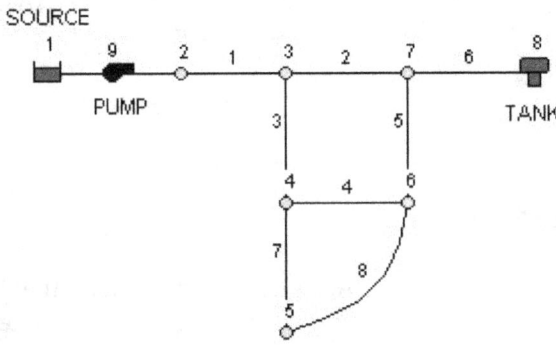

Figure 2.1 Example Pipe Network

Table 2.1 Example Network Node Properties

Node	Elevation (ft)	Demand (gpm)
1	700	0
2	700	0
3	710	150
4	700	150
5	650	200
6	700	150
7	700	0
8	830	0

Table 2.2 Example Network Pipe Properties

Pipe	Length (ft)	Diameter (inches)	C-Factor
1	3000	14	100
2	5000	12	100
3	5000	8	100
4	5000	8	100
5	5000	8	100
6	7000	10	100
7	5000	6	100
8	7000	6	100

2.3 Project Setup

Our first task is to create a new project in EPANET and make sure that certain default options are selected. To begin, launch EPANET, or if it is already running select **File >> New** (from the menu bar) to create a new project. Then select **Project >> Defaults** to open the dialog form shown in Figure 2.2. We will use this dialog to have EPANET automatically label new objects with consecutive numbers starting from 1 as they are added to the network. On the ID Labels page of the dialog, clear all of the ID Prefix fields and set the ID Increment to 1. Then select the Hydraulics page of the dialog and set the choice of Flow Units to GPM (gallons per minute). This implies that US Customary units will be used for all other quantities as well (length in feet, pipe diameter in inches, pressure in psi, etc.). Also select Hazen-Williams (H-W) as the headloss formula. If you wanted to save these choices for all future new projects you could check the **Save** box at the bottom of the form before accepting it by clicking the **OK** button.

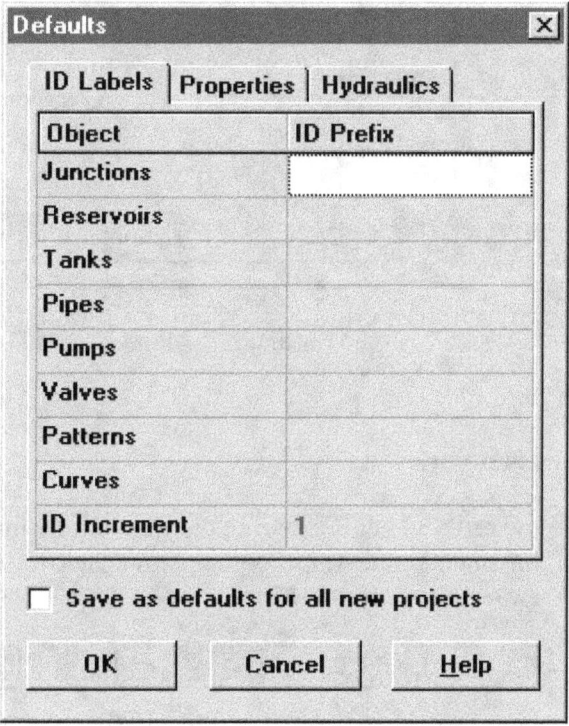

Figure 2.2 Project Defaults Dialog

Next we will select some map display options so that as we add objects to the map, we will see their ID labels and symbols displayed. Select **View >> Options** to bring up the Map Options dialog form. Select the Notation page on this form and check the settings shown in Figure 2.3 below. Then switch to the Symbols page and check all of the boxes. Click the **OK** button to accept these choices and close the dialog.

Finally, before drawing our network we should insure that our map scale settings are acceptable. Select **View >> Dimensions** to bring up the Map Dimensions dialog. Note the default dimensions assigned for a new project. These settings will suffice for this example, so click the **OK** button.

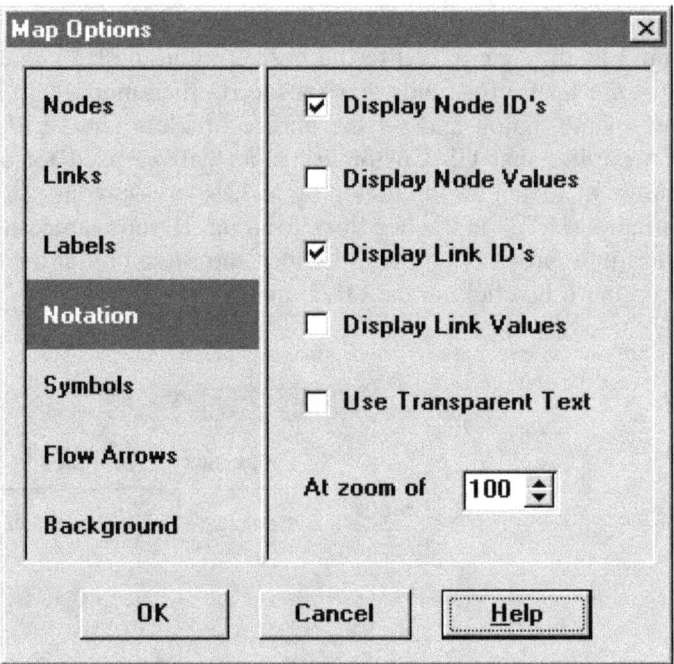

Figure 2.3 Map Options Dialog

2.4   Drawing the Network

We are now ready to begin drawing our network by making use of our mouse and the buttons contained on the Map Toolbar shown below. (If the toolbar is not visible then select **View >> Toolbars >> Map**).

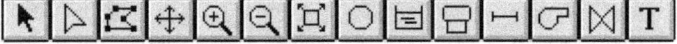

First we will add the reservoir. Click the Reservoir button . Then click the mouse on the map at the location of the reservoir (somewhere to the left of the map).

Next we will add the junction nodes. Click the Junction button and then click on the map at the locations of nodes 2 through 7.

Finally add the tank by clicking the Tank button and clicking the map where the tank is located. At this point the Network Map should look something like the drawing in Figure 2.4.

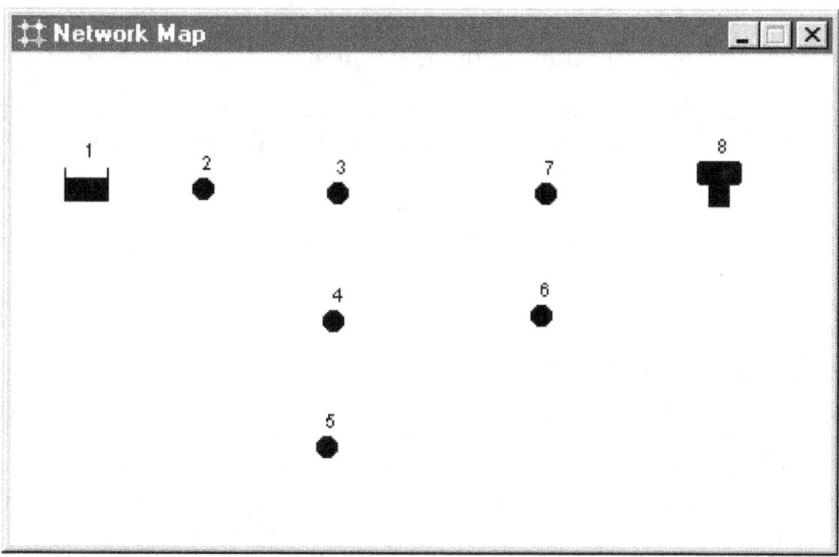

Figure 2.4 Network Map after Adding Nodes

Next we will add the pipes. Let's begin with pipe 1 connecting node 2 to node 3. First click the Pipe button ⊢⊣ on the Toolbar. Then click the mouse on node 2 on the map and then on node 3. Note how an outline of the pipe is drawn as you move the mouse from node 2 to 3. Repeat this procedure for pipes 2 through 7.

Pipe 8 is curved. To draw it, click the mouse first on Node 5. Then as you move the mouse towards Node 6, click at those points where a change of direction is needed to maintain the desired shape. Complete the process by clicking on Node 6.

Finally we will add the pump. Click the Pump button ⌖, click on node 1 and then on node 2.

Next we will label the reservoir, pump and tank. Select the Text button **T** on the Map Toolbar and click somewhere close to the reservoir (Node 1). An edit box will appear. Type in the word SOURCE and then hit the **Enter** key. Click next to the pump and enter its label, then do the same for the tank. Then click the Selection button ▸ on the Toolbar to put the map into Object Selection mode rather than Text Insertion mode.

At this point we have completed drawing the example network. Your Network Map should look like the map in Figure 2.1. If the nodes are out of position you can move them around by clicking the node to select it, and then dragging it with the left mouse button held down to its new position. Note how pipes connected to the node are moved along with the node. The labels can be repositioned in similar fashion. To re-shape the curved Pipe 8:

1. First click on Pipe 8 to select it and then click the ▷ button on the Map Toolbar to put the map into Vertex Selection mode.

2. Select a vertex point on the pipe by clicking on it and then drag it to a new position with the left mouse button held down.

3. If required, vertices can be added or deleted from the pipe by right-clicking the mouse and selecting the appropriate option from the popup menu that appears.

4. When finished, click ▣ to return to Object Selection mode.

2.5 Setting Object Properties

As objects are added to a project they are assigned a default set of properties. To change the value of a specific property for an object one must select the object into the Property Editor (Figure 2.5). There are several different ways to do this. If the Editor is already visible then you can simply click on the object or select it from the Data page of the Browser. If the Editor is not visible then you can make it appear by one of the following actions:

- Double-click the object on the map.

- Right-click on the object and select **Properties** from the pop-up menu that appears.

- Select the object from the Data page of the Browser window and then click the Browser's Edit button ▣.

Whenever the Property Editor has the focus you can press the F1 key to obtain fuller descriptions of the properties listed

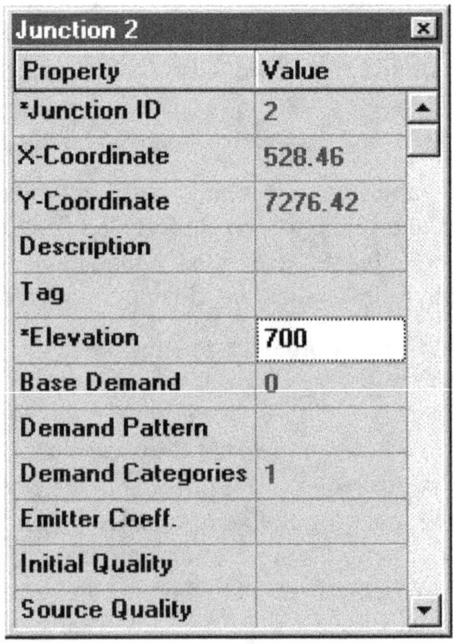

Junction 2	
Property	Value
*Junction ID	2
X-Coordinate	528.46
Y-Coordinate	7276.42
Description	
Tag	
*Elevation	700
Base Demand	0
Demand Pattern	
Demand Categories	1
Emitter Coeff.	
Initial Quality	
Source Quality	

Figure 2.5 Property Editor

Let us begin editing by selecting Node 2 into the Property Editor as shown above. We would now enter the elevation and demand for this node in the appropriate fields. You can use the **Up** and **Down** arrows on the keyboard or the mouse to move between fields. We need only click on another object (node or link) to have its properties appear next in the Property Editor. (We could also press the **Page Down** or **Page Up** key to move to the next or previous object of the same type in the database.) Thus we can simply move from object to object and fill in elevation and demand for nodes, and length, diameter, and roughness (C-factor) for links.

For the reservoir you would enter its elevation (700) in the Total Head field. For the tank, enter 830 for its elevation, 4 for its initial level, 20 for its maximum level, and 60 for its diameter. For the pump, we need to assign it a pump curve (head versus flow relationship). Enter the ID label 1 in the Pump Curve field.

Next we will create Pump Curve 1. From the Data page of the Browser window, select Curves from the dropdown list box and then click the Add button. A new Curve 1 will be added to the database and the Curve Editor dialog form will appear (see Figure 2.6). Enter the pump's design flow (600) and head (150) into this form. EPANET automatically creates a complete pump curve from this single point. The curve's equation is shown along with its shape. Click **OK** to close the Editor.

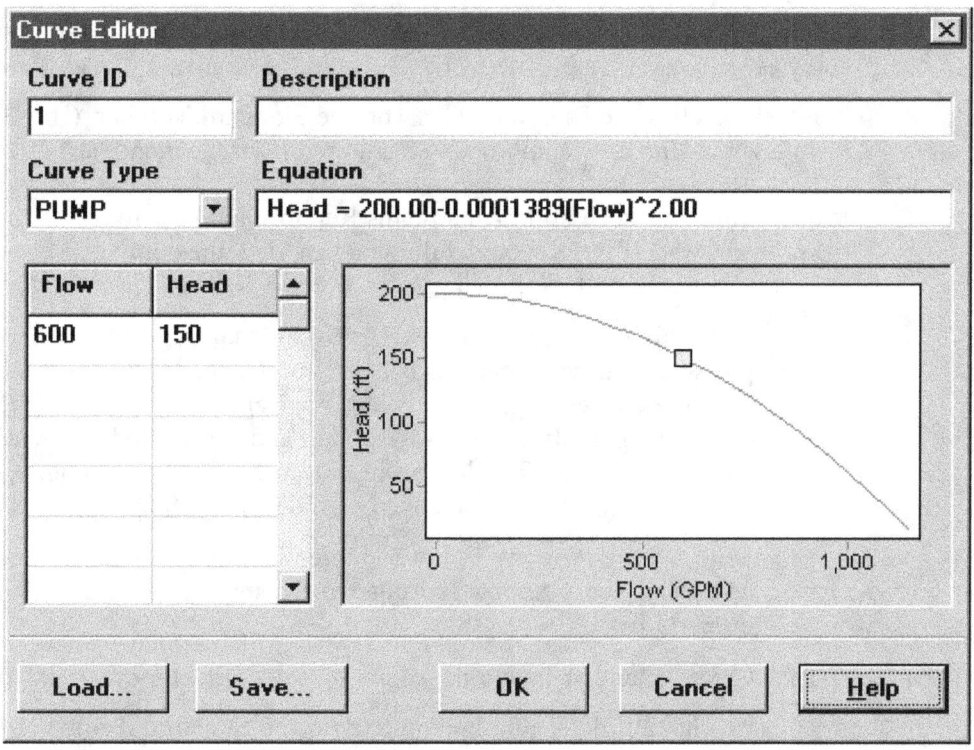

Figure 2.6 Curve Editor

2.6 Saving and Opening Projects

Having completed the initial design of our network it is a good idea to save our work to a file at this point.

1. From the **File** menu select the **Save As** option.
2. In the Save As dialog that appears, select a folder and file name under which to save this project. We suggest naming the file **tutorial.net**. (An extension of **.net** will be added to the file name if one is not supplied.)
3. Click **OK** to save the project to file.

The project data is saved to the file in a special binary format. If you wanted to save the network data to file as readable text, use the **File >> Export >> Network** command instead.

To open our project at some later time, we would select the **Open** command from the **File** menu.

2.7 Running a Single Period Analysis

We now have enough information to run a single period (or snapshot) hydraulic analysis on our example network. To run the analysis select **Project >> Run Analysis** or click the Run button on the Standard Toolbar. (If the toolbar is not visible select **View >> Toolbars >> Standard** from the menu bar).

If the run was unsuccessful then a Status Report window will appear indicating what the problem was. If it ran successfully you can view the computed results in a variety of ways. Try some of the following:

- Select Node Pressure from the Browser's Map page and observe how pressure values at the nodes become color-coded. To view the legend for the color-coding, select **View >> Legends >> Node** (or right-click on an empty portion of the map and select Node Legend from the popup menu). To change the legend intervals and colors, right-click on the legend to make the Legend Editor appear.

- Bring up the Property Editor (double-click on any node or link) and note how the computed results are displayed at the end of the property list.

- Create a tabular listing of results by selecting **Report >> Table** (or by clicking the Table button on the Standard Toolbar). Figure 2.7 displays such a table for the link results of this run. Note that flows with negative signs means that the flow is in the opposite direction to the direction in which the pipe was drawn initially.

Link ID	Flow GPM	Velocity fps	Headloss ft/Kft	Status
Pipe 1	617.42	1.29	0.80	Open
Pipe 2	382.51	1.09	0.69	Open
Pipe 3	159.91	1.02	1.00	Open
Pipe 4	29.34	0.19	0.04	Open
Pipe 5	-90.09	0.57	0.34	Open
Pipe 6	292.42	1.19	1.03	Open
Pipe 7	55.58	0.63	0.57	Open
Pipe 8	-44.42	0.50	0.38	Open

Figure 2.7 Example Table of Link Results

2.8 Running an Extended Period Analysis

To make our network more realistic for analyzing an extended period of operation we will create a Time Pattern that makes demands at the nodes vary in a periodic way over the course of a day. For this simple example we will use a pattern time step of 6 hours thus making demands change at four different times of the day. (A 1-hour pattern time step is a more typical number and is the default assigned to new projects.) We set the pattern time step by selecting Options-Times from the Data Browser, clicking the Browser's Edit button to make the Property Editor appear (if its not already visible), and entering 6 for the value of the Pattern Time Step (as shown in Figure 2.8 below). While we have the Time Options available we can also set the duration for which we want the extended period to run. Let's use a 3-day period of time (enter 72 hours for the Duration property).

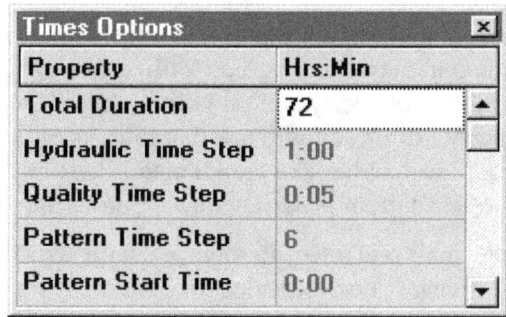

Figure 2.8 Times Options

To create the pattern, select the Patterns category in the Browser and then click the Add button . A new Pattern 1 will be created and the Pattern Editor dialog should appear (see Figure 2.9). Enter the multiplier values 0.5, 1.3, 1.0, 1.2 for the time periods 1 to 4 that will give our pattern a duration of 24 hours. The multipliers are used to modify the demand from its base level in each time period. Since we are making a run of 72 hours, the pattern will wrap around to the start after each 24-hour interval of time.

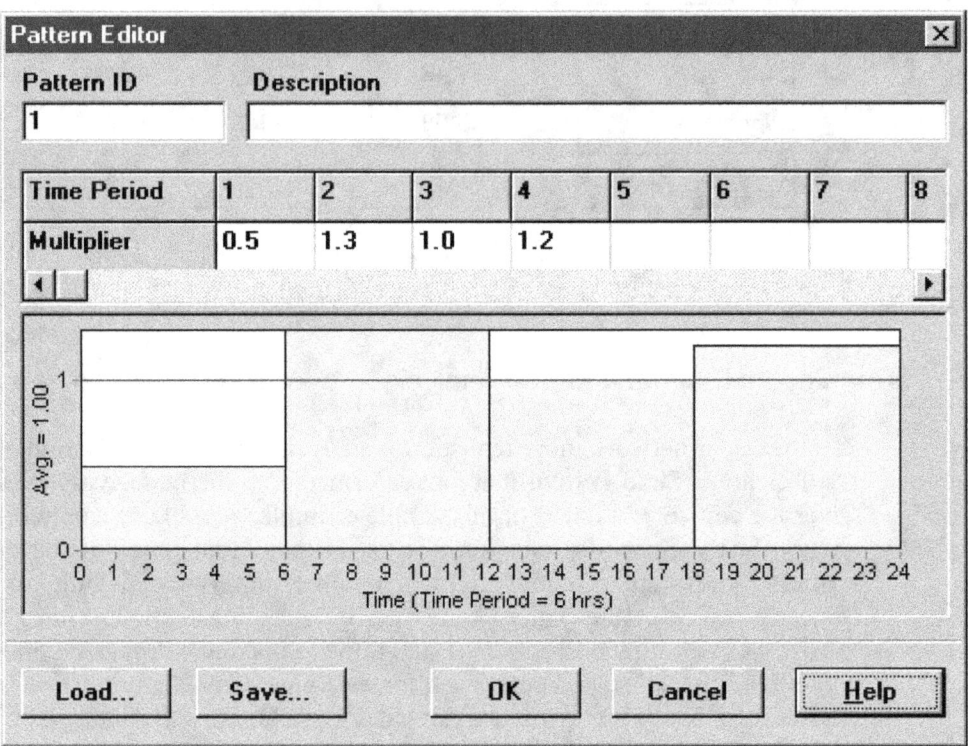

Figure 2.9 Pattern Editor

We now need to assign Pattern 1 to the Demand Pattern property of all of the junctions in our network. We can utilize one of EPANET's Hydraulic Options to avoid having to edit each junction individually. If you bring up the Hydraulic Options in the Property Editor you will see that there is an item called Default Pattern. Setting its value equal to 1 will make the Demand Pattern at each junction equal Pattern 1, as long as no other pattern is assigned to the junction.

Next run the analysis (select **Project >> Run Analysis** or click the button on the Standard Toolbar). For extended period analysis you have several more ways in which to view results:

- The scrollbar in the Browser's Time controls is used to display the network map at different points in time. Try doing this with Pressure selected as the node parameter and Flow as the link parameter.

- The VCR-style buttons in the Browser can animate the map through time. Click the Forward button ▶ to start the animation and the Stop button ■ to stop it.

- Add flow direction arrows to the map (select **View >> Options**, select the Flow Arrows page from the Map Options dialog, and check a style of arrow that you wish to use). Then begin the animation again and note the change in flow direction through the pipe connected to the tank as the tank fills and empties over time.

- Create a time series plot for any node or link. For example, to see how the water elevation in the tank changes with time:

    1. Click on the tank.

    2. Select **Report >> Graph** (or click the Graph button on the Standard Toolbar) which will display a Graph Selection dialog box.

    3. Select the Time Series button on the dialog.

    4. Select Head as the parameter to plot.

    5. Click **OK** to accept your choice of graph.

Note the periodic behavior of the water elevation in the tank over time (Figure 2.10).

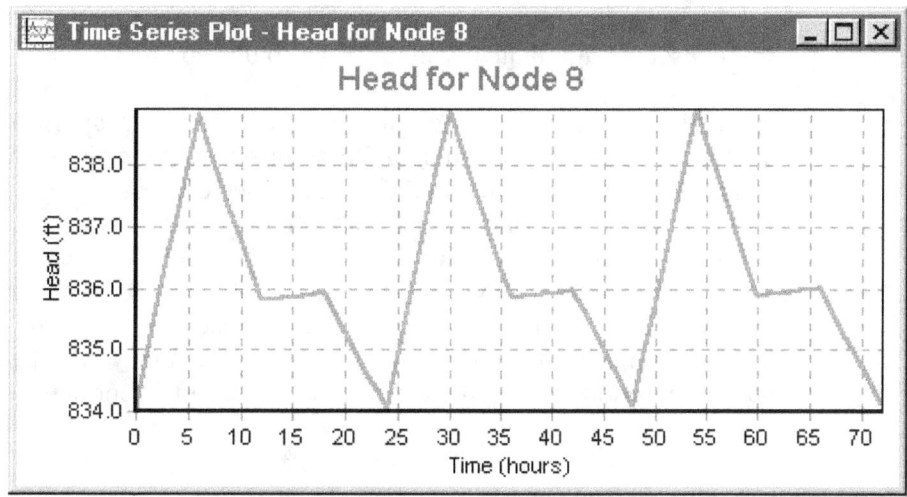

Figure 2.10 Example Time Series Plot

2.9 Running a Water Quality Analysis

Next we show how to extend the analysis of our example network to include water quality. The simplest case would be tracking the growth in water age throughout the network over time. To make this analysis we only have to select Age for the Parameter property in the Quality Options (select Options-Quality from the Data page of the Browser, then click the Browser's Edit button to make the Property Editor appear). Run the analysis and select Age as the parameter to view on the map. Create a time series plot for Age in the tank. Note that unlike water level, 72 hours is not enough time for the tank to reach periodic behavior for water age. (The default initial condition is to start all nodes with an age of 0.) Try repeating the simulation using a 240-hour duration or assigning an initial age of 60 hours to the tank (enter 60 as the value of Initial Quality in the Property Editor for the tank).

Finally we show how to simulate the transport and decay of chlorine through the network. Make the following changes to the database:

1. Select Options-Quality to edit from the Data Browser. In the Property Editor's Parameter field type in the word Chlorine.

2. Switch to Options-Reactions in the Browser. For Global Bulk Coefficient enter a value of -1.0. This reflects the rate at which chlorine will decay due to reactions in the bulk flow over time. This rate will apply to all pipes in the network. You could edit this value for individual pipes if you needed to.

3. Click on the reservoir node and set its Initial Quality to 1.0. This will be the concentration of chlorine that continuously enters the network. (Reset the initial quality in the Tank to 0 if you had changed it.)

Now run the example. Use the Time controls on the Map Browser to see how chlorine levels change by location and time throughout the simulation. Note how for this simple network, only junctions 5, 6, and 7 see depressed chlorine levels because of being fed by low chlorine water from the tank. Create a reaction report for this run by selecting **Report >> Reaction** from the main menu. The report should look like Figure 2.11. It shows on average how much chlorine loss occurs in the pipes as opposed to the tank. The term "bulk" refers to reactions occurring in the bulk fluid while "wall" refers to reactions with material on the pipe wall. The latter reaction is zero because we did not specify any wall reaction coefficient in this example.

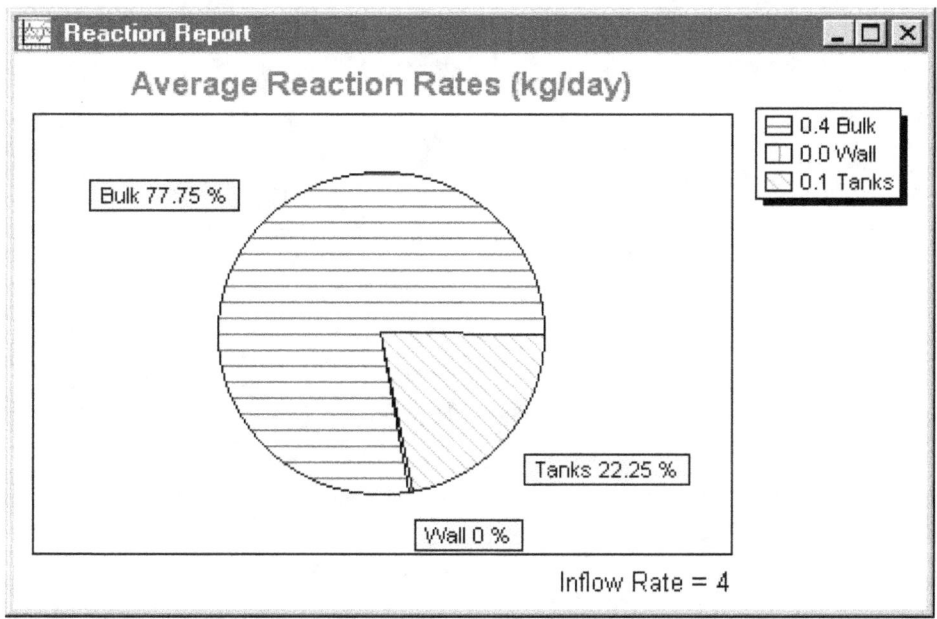

Figure 2.11 Example Reaction Report

We have only touched the surface of the various capabilities offered by EPANET. Some additional features of the program that you should experiment with are:

- Editing a property for a group of objects that lie within a user-defined area.
- Using Control statements to base pump operation on time of day or tank water levels.
- Exploring different Map Options, such as making node size be related to value.
- Attaching a backdrop map (such as a street map) to the network map.
- Creating different types of graphs, such as profile plots and contour plots.
- Adding calibration data to a project and viewing a calibration report.
- Copying the map, a graph, or a report to the clipboard or to a file.
- Saving and retrieving a design scenario (i.e., current nodal demands, pipe roughness values, etc.).

(This page intentionally left blank.)

CHAPTER 3 - THE NETWORK MODEL

This chapter discusses how EPANET models the physical objects that constitute a distribution system as well as its operational parameters. Details about how this information is entered into the program are presented in later chapters. An overview is also given on the computational methods that EPANET uses to simulate hydraulic and water quality transport behavior.

3.1 Physical Components

EPANET models a water distribution system as a collection of links connected to nodes. The links represent pipes, pumps, and control valves. The nodes represent junctions, tanks, and reservoirs. The figure below illustrates how these objects can be connected to one another to form a network.

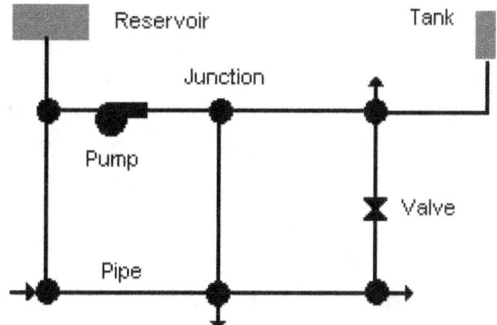

Figure 3.1 Physical Components in a Water Distribution System

Junctions

Junctions are points in the network where links join together and where water enters or leaves the network. The basic input data required for junctions are:

- elevation above some reference (usually mean sea level)
- water demand (rate of withdrawal from the network)
- initial water quality.

The output results computed for junctions at all time periods of a simulation are:

- hydraulic head (internal energy per unit weight of fluid)
- pressure
- water quality.

Junctions can also:

- have their demand vary with time
- have multiple categories of demands assigned to them
- have negative demands indicating that water is entering the network
- be water quality sources where constituents enter the network
- contain emitters (or sprinklers) which make the outflow rate depend on the pressure.

Reservoirs

Reservoirs are nodes that represent an infinite external source or sink of water to the network. They are used to model such things as lakes, rivers, groundwater aquifers, and tie-ins to other systems. Reservoirs can also serve as water quality source points.

The primary input properties for a reservoir are its hydraulic head (equal to the water surface elevation if the reservoir is not under pressure) and its initial quality for water quality analysis.

Because a reservoir is a boundary point to a network, its head and water quality cannot be affected by what happens within the network. Therefore it has no computed output properties. However its head can be made to vary with time by assigning a time pattern to it (see Time Patterns below).

Tanks

Tanks are nodes with storage capacity, where the volume of stored water can vary with time during a simulation. The primary input properties for tanks are:

- bottom elevation (where water level is zero)
- diameter (or shape if non-cylindrical)
- initial, minimum and maximum water levels
- initial water quality.

The principal outputs computed over time are:

- hydraulic head (water surface elevation)
- water quality.

Tanks are required to operate within their minimum and maximum levels. EPANET stops outflow if a tank is at its minimum level and stops inflow if it is at its maximum level. Tanks can also serve as water quality source points.

Emitters

Emitters are devices associated with junctions that model the flow through a nozzle or orifice that discharges to the atmosphere. The flow rate through the emitter varies as a function of the pressure available at the node:

$$q = C\, p^{\gamma}$$

where q = flow rate, p = pressure, C = discharge coefficient, and γ = pressure exponent. For nozzles and sprinkler heads γ equals 0.5 and the manufacturer usually provides the value of the discharge coefficient in units of gpm/psi$^{0.5}$ (stated as the flow through the device at a 1 psi pressure drop).

Emitters are used to model flow through sprinkler systems and irrigation networks. They can also be used to simulate leakage in a pipe connected to the junction (if a discharge coefficient and pressure exponent for the leaking crack or joint can be estimated) or compute a fire flow at the junction (the flow available at some minimum residual pressure). In the latter case one would use a very high value of the discharge coefficient (e.g., 100 times the maximum flow expected) and modify the junction's elevation to include the equivalent head of the pressure target. EPANET treats emitters as a property of a junction and not as a separate network component.

Pipes

Pipes are links that convey water from one point in the network to another. EPANET assumes that all pipes are full at all times. Flow direction is from the end at higher hydraulic head (internal energy per weight of water) to that at lower head. The principal hydraulic input parameters for pipes are:

- start and end nodes
- diameter
- length
- roughness coefficient (for determining headloss)
- status (open, closed, or contains a check valve).

The status parameter allows pipes to implicitly contain shutoff (gate) valves and check (non-return) valves (which allow flow in only one direction).

The water quality inputs for pipes consist of:

- bulk reaction coefficient
- wall reaction coefficient.

These coefficients are explained more thoroughly in Section 3.4 below.

Computed outputs for pipes include:
- flow rate
- velocity
- headloss
- Darcy-Weisbach friction factor
- average reaction rate (over the pipe length)
- average water quality (over the pipe length).

The hydraulic head lost by water flowing in a pipe due to friction with the pipe walls can be computed using one of three different formulas:
- Hazen-Williams formula
- Darcy-Weisbach formula
- Chezy-Manning formula

The Hazen-Williams formula is the most commonly used headloss formula in the US. It cannot be used for liquids other than water and was originally developed for turbulent flow only. The Darcy-Weisbach formula is the most theoretically correct. It applies over all flow regimes and to all liquids. The Chezy-Manning formula is more commonly used for open channel flow.

Each formula uses the following equation to compute headloss between the start and end node of the pipe:

$$h_L = Aq^B$$

where h_L = headloss (Length), q = flow rate (Volume/Time), A = resistance coefficient, and B = flow exponent. Table 3.1 lists expressions for the resistance coefficient and values for the flow exponent for each of the formulas. Each formula uses a different pipe roughness coefficient that must be determined empirically. Table 3.2 lists general ranges of these coefficients for different types of new pipe materials. Be aware that a pipe's roughness coefficient can change considerably with age.

With the Darcy-Weisbach formula EPANET uses different methods to compute the friction factor f depending on the flow regime:
- The Hagen–Poiseuille formula is used for laminar flow (Re < 2,000).
- The Swamee and Jain approximation to the Colebrook-White equation is used for fully turbulent flow (Re > 4,000).
- A cubic interpolation from the Moody Diagram is used for transitional flow (2,000 < Re < 4,000).

Consult Appendix D for the actual equations used.

Table 3.1 Pipe Headloss Formulas for Full Flow
(for headloss in feet and flow rate in cfs)

Formula	Resistance Coefficient (A)	Flow Exponent (B)
Hazen-Williams	$4.727\, C^{-1.852}\, d^{-4.871}\, L$	1.852
Darcy-Weisbach	$0.0252\, f(\varepsilon,d,q)\, d^{-5}\, L$	2
Chezy-Manning	$4.66\, n^2\, d^{-5.33}\, L$	2

Notes: C = Hazen-Williams roughness coefficient
ε = Darcy-Weisbach roughness coefficient (ft)
f = friction factor (dependent on ε, d, and q)
n = Manning roughness coefficient
d = pipe diameter (ft)
L = pipe length (ft)
q = flow rate (cfs)

Table 3.2 Roughness Coefficients for New Pipe

Material	Hazen-Williams C (unitless)	Darcy-Weisbach ε (feet x 10^{-3})	Manning's n (unitless)
Cast Iron	130 – 140	0.85	0.012 - 0.015
Concrete or Concrete Lined	120 – 140	1.0 - 10	0.012 - 0.017
Galvanized Iron	120	0.5	0.015 - 0.017
Plastic	140 – 150	0.005	0.011 - 0.015
Steel	140 – 150	0.15	0.015 - 0.017
Vitrified Clay	110		0.013 - 0.015

Pipes can be set open or closed at preset times or when specific conditions exist, such as when tank levels fall below or above certain set points, or when nodal pressures fall below or above certain values. See the discussion of Controls in Section 3.2.

Minor Losses

Minor head losses (also called local losses) are caused by the added turbulence that occurs at bends and fittings. The importance of including such losses depends on the layout of the network and the degree of accuracy required. They can be accounted for by assigning the pipe a minor loss coefficient. The minor headloss becomes the product of this coefficient and the velocity head of the pipe, i.e.,

$$h_L = K\left(\frac{v^2}{2g}\right)$$

where K = minor loss coefficient, v = flow velocity (Length/Time), and g = acceleration of gravity (Length/Time^2). Table 3.3 provides minor loss coefficients for several types of fittings.

Table 3.3 Minor Loss Coefficients for Selected Fittings

FITTING	LOSS COEFFICIENT
Globe valve, fully open	10.0
Angle valve, fully open	5.0
Swing check valve, fully open	2.5
Gate valve, fully open	0.2
Short-radius elbow	0.9
Medium-radius elbow	0.8
Long-radius elbow	0.6
45 degree elbow	0.4
Closed return bend	2.2
Standard tee - flow through run	0.6
Standard tee - flow through branch	1.8
Square entrance	0.5
Exit	1.0

Pumps

Pumps are links that impart energy to a fluid thereby raising its hydraulic head. The principal input parameters for a pump are its start and end nodes and its pump curve (the combination of heads and flows that the pump can produce). In lieu of a pump curve, the pump could be represented as a constant energy device, one that supplies a constant amount of energy (horsepower or kilowatts) to the fluid for all combinations of flow and head.

The principal output parameters are flow and head gain. Flow through a pump is unidirectional and EPANET will not allow a pump to operate outside the range of its pump curve.

Variable speed pumps can also be considered by specifying that their speed setting be changed under these same types of conditions. By definition, the original pump curve supplied to the program has a relative speed setting of 1. If the pump speed doubles, then the relative setting would be 2; if run at half speed, the relative setting is 0.5 and so on. Changing the pump speed shifts the position and shape of the pump curve (see the section on Pump Curves below).

As with pipes, pumps can be turned on and off at preset times or when certain conditions exist in the network. A pump's operation can also be described by assigning it a time pattern of relative speed settings. EPANET can also compute the energy consumption and cost of a pump. Each pump can be assigned an efficiency curve and schedule of energy prices. If these are not supplied then a set of global energy options will be used.

Flow through a pump is unidirectional. If system conditions require more head than the pump can produce, EPANET shuts the pump off. If more than the maximum flow is required, EPANET extrapolates the pump curve to the required flow, even if this produces a negative head. In both cases a warning message will be issued.

Valves

Valves are links that limit the pressure or flow at a specific point in the network. Their principal input parameters include:

- start and end nodes
- diameter
- setting
- status.

The computed outputs for a valve are flow rate and headloss.

The different types of valves included in EPANET are:

- Pressure Reducing Valve (PRV)
- Pressure Sustaining Valve (PSV)
- Pressure Breaker Valve (PBV)
- Flow Control Valve (FCV)
- Throttle Control Valve (TCV)
- General Purpose Valve (GPV).

PRVs limit the pressure at a point in the pipe network. EPANET computes in which of three different states a PRV can be in:

- partially opened (i.e., active) to achieve its pressure setting on its downstream side when the upstream pressure is above the setting
- fully open if the upstream pressure is below the setting
- closed if the pressure on the downstream side exceeds that on the upstream side (i.e., reverse flow is not allowed).

PSVs maintain a set pressure at a specific point in the pipe network. EPANET computes in which of three different states a PSV can be in:

- partially opened (i.e., active) to maintain its pressure setting on its upstream side when the downstream pressure is below this value
- fully open if the downstream pressure is above the setting

- closed if the pressure on the downstream side exceeds that on the upstream side (i.e., reverse flow is not allowed).

PBVs force a specified pressure loss to occur across the valve. Flow through the valve can be in either direction. PBV's are not true physical devices but can be used to model situations where a particular pressure drop is known to exist.

FCVs limit the flow to a specified amount. The program produces a warning message if this flow cannot be maintained without having to add additional head at the valve (i.e., the flow cannot be maintained even with the valve fully open).

TCVs simulate a partially closed valve by adjusting the minor head loss coefficient of the valve. A relationship between the degree to which a valve is closed and the resulting head loss coefficient is usually available from the valve manufacturer.

GPVs are used to represent a link where the user supplies a special flow - head loss relationship instead of following one of the standard hydraulic formulas. They can be used to model turbines, well draw-down or reduced-flow backflow prevention valves.

Shutoff (gate) valves and check (non-return) valves, which completely open or close pipes, are not considered as separate valve links but are instead included as a property of the pipe in which they are placed.

Each type of valve has a different type of setting parameter that describes its operating point (pressure for PRVs, PSVs, and PBVs; flow for FCVs; loss coefficient for TCVs, and head loss curve for GPVs).

Valves can have their control status overridden by specifying they be either completely open or completely closed. A valve's status and its setting can be changed during the simulation by using control statements.

Because of the ways in which valves are modeled the following rules apply when adding valves to a network:
- a PRV, PSV or FCV cannot be directly connected to a reservoir or tank (use a length of pipe to separate the two)
- PRVs cannot share the same downstream node or be linked in series
- two PSVs cannot share the same upstream node or be linked in series
- a PSV cannot be connected to the downstream node of a PRV.

3.2 Non-Physical Components

In addition to physical components, EPANET employs three types of informational objects – curves, patterns, and controls - that describe the behavior and operational aspects of a distribution system.

Curves

Curves are objects that contain data pairs representing a relationship between two quantities. Two or more objects can share the same curve. An EPANET model can utilize the following types of curves:

- Pump Curve
- Efficiency Curve
- Volume Curve
- Head Loss Curve

Pump Curve

A Pump Curve represents the relationship between the head and flow rate that a pump can deliver at its nominal speed setting. Head is the head gain imparted to the water by the pump and is plotted on the vertical (Y) axis of the curve in feet (meters). Flow rate is plotted on the horizontal (X) axis in flow units. A valid pump curve must have decreasing head with increasing flow.

EPANET will use a different shape of pump curve depending on the number of points supplied (see Figure 3.2):

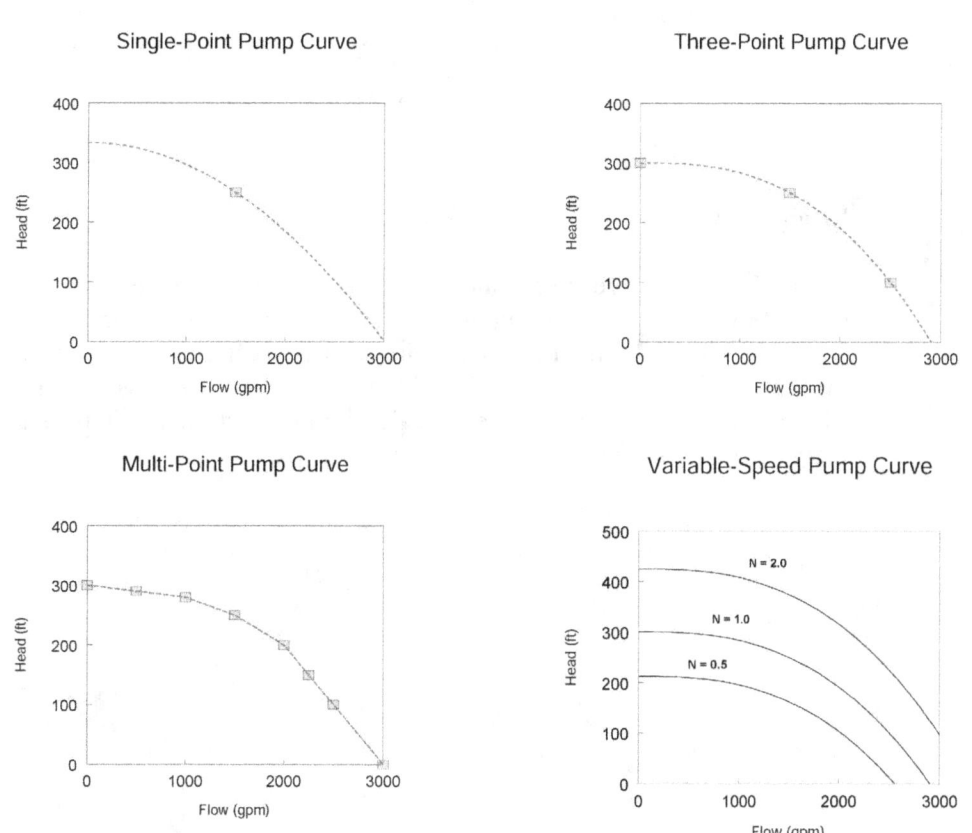

Figure 3.2 Example Pump Curves

Single-Point Curve - A single-point pump curve is defined by a single head-flow combination that represents a pump's desired operating point. EPANET adds two more points to the curve by assuming a shutoff head at zero flow equal to 133% of the design head and a maximum flow at zero head equal to twice the design flow. It then treats the curve as a three-point curve.

Three-Point Curve - A three-point pump curve is defined by three operating points: a Low Flow point (flow and head at low or zero flow condition), a Design Flow point (flow and head at desired operating point), and a Maximum Flow point (flow and head at maximum flow). EPANET tries to fit a continuous function of the form

$$h_G = A - Bq^C$$

through the three points to define the entire pump curve. In this function, h_g = head gain, q = flow rate, and A, B, and C are constants.

Multi-Point Curve – A multi-point pump curve is defined by providing either a pair of head-flow points or four or more such points. EPANET creates a complete curve by connecting the points with straight-line segments.

For variable speed pumps, the pump curve shifts as the speed changes. The relationships between flow (Q) and head (H) at speeds N1 and N2 are

$$\frac{Q_1}{Q_2} = \frac{N_1}{N_2} \qquad \frac{H_1}{H_2} = \left(\frac{N_1}{N_2}\right)^2$$

Efficiency Curve

An Efficiency Curve determines pump efficiency (Y in percent) as a function of pump flow rate (X in flow units). An example efficiency curve is shown in Figure 3.3. Efficiency should represent wire-to-water efficiency that takes into account mechanical losses in the pump itself as well as electrical losses in the pump's motor. The curve is used only for energy calculations. If not supplied for a specific pump then a fixed global pump efficiency will be used.

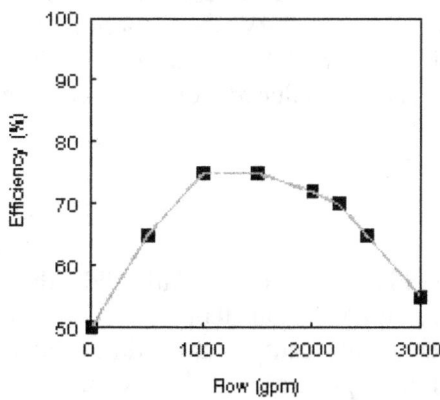

Figure 3.3 Pump Efficiency Curve

Volume Curve

A Volume Curve determines how storage tank volume (Y in cubic feet or cubic meters) varies as a function of water level (X in feet or meters). It is used when it is necessary to accurately represent tanks whose cross-sectional area varies with height. The lower and upper water levels supplied for the curve must contain the lower and upper levels between which the tank operates. An example of a tank volume curve is given below.

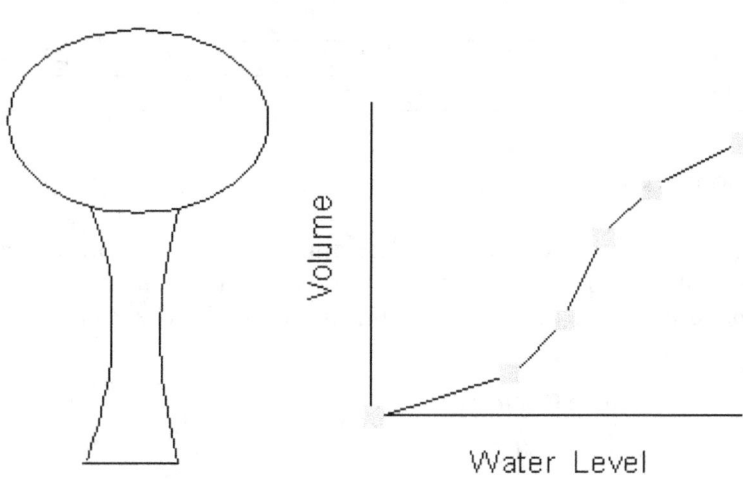

Figure 3.4 Tank Volume Curve

Headloss Curve

A Headloss Curve is used to described the headloss (Y in feet or meters) through a General Purpose Valve (GPV) as a function of flow rate (X in flow units). It provides the capability to model devices and situations with unique headloss-flow relationships, such as reduced flow - backflow prevention valves, turbines, and well draw-down behavior.

Time Patterns

A Time Pattern is a collection of multipliers that can be applied to a quantity to allow it to vary over time. Nodal demands, reservoir heads, pump schedules, and water quality source inputs can all have time patterns associated with them. The time interval used in all patterns is a fixed value, set with the project's Time Options (see Section 8.1). Within this interval a quantity remains at a constant level, equal to the product of its nominal value and the pattern's multiplier for that time period. Although all time patterns must utilize the same time interval, each can have a different number of periods. When the simulation clock exceeds the number of periods in a pattern, the pattern wraps around to its first period again.

As an example of how time patterns work consider a junction node with an average demand of 10 GPM. Assume that the time pattern interval has been set to 4 hours and a pattern with the following multipliers has been specified for demand at this node:

Period	1	2	3	4	5	6
Multiplier	0.5	0.8	1.0	1.2	0.9	0.7

Then during the simulation the actual demand exerted at this node will be as follows:

Hours	0-4	4-8	8-12	12-16	16-20	20-24	24-28
Demand	5	8	10	12	9	7	5

Controls

Controls are statements that determine how the network is operated over time. They specify the status of selected links as a function of time, tank water levels, and pressures at select points within the network. There are two categories of controls that can be used:

- Simple Controls
- Rule-Based Controls

Simple Controls

Simple controls change the status or setting of a link based on:
- the water level in a tank,
- the pressure at a junction,
- the time into the simulation,

- the time of day.

They are statements expressed in one of the following three formats:

```
LINK x status IF NODE y ABOVE/BELOW z
LINK x status AT TIME t
LINK x status AT CLOCKTIME c AM/PM
```

where:

x	=	a link ID label,
$status$	=	OPEN or CLOSED, a pump speed setting, or a control valve setting,
y	=	a node ID label,
z	=	a pressure for a junction or a water level for a tank,
t	=	a time since the start of the simulation in decimal hours or in hours:minutes notation,
c	=	a 24-hour clock time.

Some examples of simple controls are:

Control Statement	Meaning
`LINK 12 CLOSED IF NODE 23 ABOVE 20`	(Close Link 12 when the level in Tank 23 exceeds 20 ft.)
`LINK 12 OPEN IF NODE 130 BELOW 30`	(Open Link 12 if the pressure at Node 130 drops below 30 psi)
`LINK 12 1.5 AT TIME 16`	(Set the relative speed of pump 12 to 1.5 at 16 hours into the simulation)
`LINK 12 CLOSED AT CLOCKTIME 10 AM` `LINK 12 OPEN AT CLOCKTIME 8 PM`	(Link 12 is repeatedly closed at 10 AM and opened at 8 PM throughout the simulation)

There is no limit on the number of simple control statements that can be used.

Note: Level controls are stated in terms of the height of water above the tank bottom, not the elevation (total head) of the water surface.

Note: Using a pair of pressure controls to open and close a link can cause the system to become unstable if the pressure settings are too close to one another. In this case using a pair of Rule-Based controls might provide more stability.

Rule-Based Controls

Rule-Based Controls allow link status and settings to be based on a combination of conditions that might exist in the network after an initial hydraulic state of the system is computed. Here are several examples of Rule-Based Controls:

Example 1:
This set of rules shuts down a pump and opens a by-pass pipe when the level in a tank exceeds a certain value and does the opposite when the level is below another value.

```
RULE 1
IF   TANK   1 LEVEL ABOVE 19.1
THEN PUMP 335 STATUS IS CLOSED
AND  PIPE 330 STATUS IS OPEN

RULE 2
IF   TANK   1 LEVEL BELOW 17.1
THEN PUMP 335 STATUS IS OPEN
AND  PIPE 330 STATUS IS CLOSED
```

Example 2:
These rules change the tank level at which a pump turns on depending on the time of day.

```
RULE 3
IF   SYSTEM CLOCKTIME >= 8 AM
AND  SYSTEM CLOCKTIME <  6 PM
AND  TANK 1 LEVEL BELOW 12
THEN PUMP 335 STATUS IS OPEN

RULE 4
IF   SYSTEM CLOCKTIME >= 6 PM
OR   SYSTEM CLOCKTIME <  8 AM
AND  TANK 1 LEVEL BELOW 14
THEN PUMP 335 STATUS IS OPEN
```

A description of the formats used with Rule-Based controls can be found in Appendix C, under the [RULES] heading (page 150).

3.3 Hydraulic Simulation Model

EPANET's hydraulic simulation model computes junction heads and link flows for a fixed set of reservoir levels, tank levels, and water demands over a succession of points in time. From one time step to the next reservoir levels and junction demands are updated according to their prescribed time patterns while tank levels are updated using the current flow solution. The solution for heads and flows at a particular point in time involves solving simultaneously the conservation of flow equation for each junction and the headloss relationship across each link in the network. This process, known as "hydraulically balancing" the network, requires using an iterative technique to solve the nonlinear equations involved. EPANET employs the "Gradient Algorithm" for this purpose. Consult Appendix D for details.

The hydraulic time step used for extended period simulation (EPS) can be set by the user. A typical value is 1 hour. Shorter time steps than normal will occur automatically whenever one of the following events occurs:

- the next output reporting time period occurs
- the next time pattern period occurs
- a tank becomes empty or full
- a simple control or rule-based control is activated.

3.4 Water Quality Simulation Model

Basic Transport

EPANET's water quality simulator uses a Lagrangian time-based approach to track the fate of discrete parcels of water as they move along pipes and mix together at junctions between fixed-length time steps. These water quality time steps are typically much shorter than the hydraulic time step (e.g., minutes rather than hours) to accommodate the short times of travel that can occur within pipes.

The method tracks the concentration and size of a series of non-overlapping segments of water that fills each link of the network. As time progresses, the size of the most upstream segment in a link increases as water enters the link while an equal loss in size of the most downstream segment occurs as water leaves the link. The size of the segments in between these remains unchanged.

For each water quality time step, the contents of each segment are subjected to reaction, a cumulative account is kept of the total mass and flow volume entering each node, and the positions of the segments are updated. New node concentrations are then calculated, which include the contributions from any external sources. Storage tank concentrations are updated depending on the type of mixing model that is used (see below). Finally, a new segment will be created at the end of each link that receives inflow from a node if the new node quality differs by a user-specified tolerance from that of the link's last segment.

Initially each pipe in the network consists of a single segment whose quality equals the initial quality assigned to the upstream node. Whenever there is a flow reversal in a pipe, the pipe's parcels are re-ordered from front to back.

Mixing in Storage Tanks

EPANET can use four different types of models to characterize mixing within storage tanks as illustrated in Figure 3.5:

- Complete Mixing
- Two-Compartment Mixing
- FIFO Plug Flow
- LIFO Plug Flow

Different models can be used with different tanks within a network.

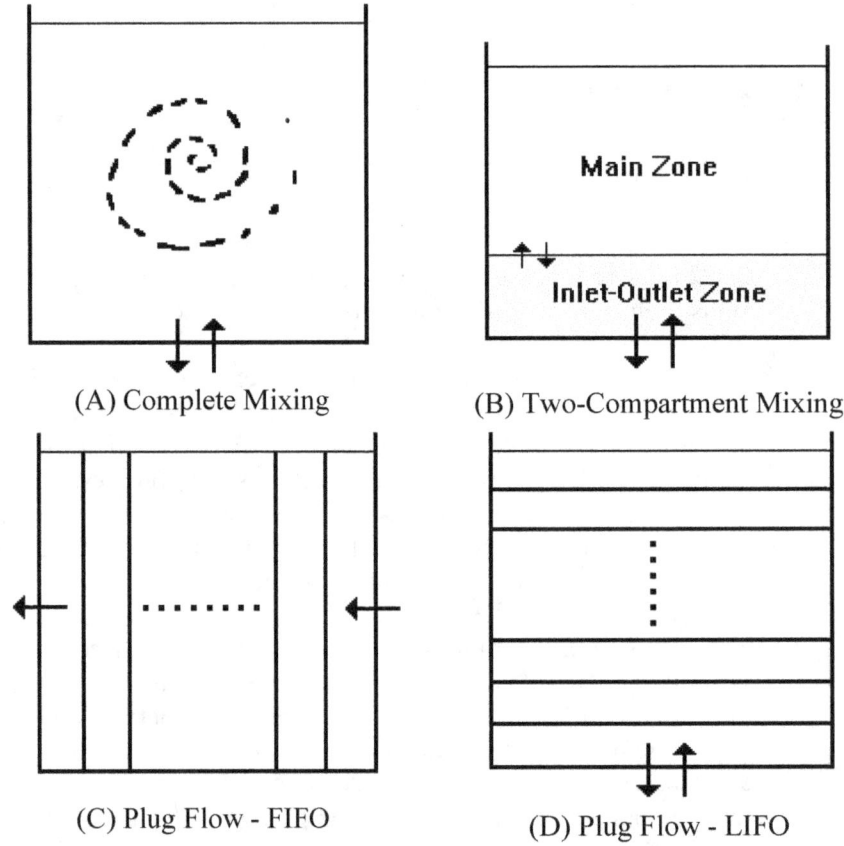

Figure 3.5 Tank Mixing Models

The Complete Mixing model (Figure 3.5(a)) assumes that all water that enters a tank is instantaneously and completely mixed with the water already in the tank. It is the simplest form of mixing behavior to assume, requires no extra parameters to describe it, and seems to apply quite well to a large number of facilities that operate in fill-and-draw fashion.

The Two-Compartment Mixing model (Figure 3.5(b)) divides the available storage volume in a tank into two compartments, both of which are assumed completely mixed. The inlet/outlet pipes of the tank are assumed to be located in the first compartment. New water that enters the tank mixes with the water in the first compartment. If this compartment is full, then it sends its overflow to the second compartment where it completely mixes with the water already stored there. When water leaves the tank, it exits from the first compartment, which if full, receives an equivalent amount of water from the second compartment to make up the difference. The first compartment is capable of simulating short-circuiting between inflow and outflow while the second compartment can represent dead zones. The user must supply a single parameter, which is the fraction of the total tank volume devoted to the first compartment.

The FIFO Plug Flow model (Figure 3.5(c)) assumes that there is no mixing of water at all during its residence time in a tank. Water parcels move through the tank in a segregated fashion where the first parcel to enter is also the first to leave. Physically speaking, this model is most appropriate for baffled tanks that operate with simultaneous inflow and outflow. There are no additional parameters needed to describe this mixing model.

The LIFO Plug Flow model (Figure 3.5(d)) also assumes that there is no mixing between parcels of water that enter a tank. However in contrast to FIFO Plug Flow, the water parcels stack up one on top of another, where water enters and leaves the tank on the bottom. This type of model might apply to a tall, narrow standpipe with an inlet/outlet pipe at the bottom and a low momentum inflow. It requires no additional parameters be provided.

Water Quality Reactions

EPANET can track the growth or decay of a substance by reaction as it travels through a distribution system. In order to do this it needs to know the rate at which the substance reacts and how this rate might depend on substance concentration. Reactions can occur both within the bulk flow and with material along the pipe wall. This is illustrated in Figure 3.6. In this example free chlorine (HOCl) is shown reacting with natural organic matter (NOM) in the bulk phase and is also transported through a boundary layer at the pipe wall to oxidize iron (Fe) released from pipe wall corrosion. Bulk fluid reactions can also occur within tanks. EPANET allows a modeler to treat these two reaction zones separately.

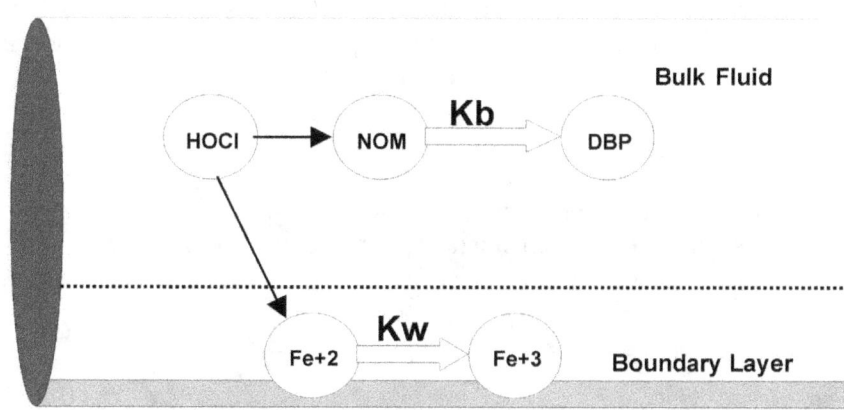

Figure 3.6 Reaction Zones Within a Pipe

Bulk Reactions

EPANET models reactions occurring in the bulk flow with n-th order kinetics, where the instantaneous rate of reaction (R in mass/volume/time) is assumed to be concentration-dependent according to

$$R = K_b C^n$$

Here K_b = a bulk reaction rate coefficient, C = reactant concentration (mass/volume), and n = a reaction order. K_b has units of concentration raised to the (1-n) power divided by time. It is positive for growth reactions and negative for decay reactions.

EPANET can also consider reactions where a limiting concentration exists on the ultimate growth or loss of the substance. In this case the rate expression becomes

$$R = K_b (C_L - C) C^{(n-1)} \quad \text{for } n > 0, K_b > 0$$
$$R = K_b (C - C_L) C^{(n-1)} \quad \text{for } n > 0, K_b < 0$$

where C_L = the limiting concentration. Thus there are three parameters (K_b, C_L, and n) that are used to characterize bulk reaction rates. Some special cases of well-known kinetic models include the following (See Appendix D for more examples):

Model	Parameters	Examples
First-Order Decay	$C_L = 0$, $K_b < 0$, $n = 1$	Chlorine
First-Order Saturation Growth	$C_L > 0$, $K_b > 0$, $n = 1$	Trihalomethanes
Zero-Order Kinetics	$C_L = 0$, $K_b \neq 0$, $n = 0$	Water Age
No Reaction	$C_L = 0$, $K_b = 0$	Fluoride Tracer

The K_b for first-order reactions can be estimated by placing a sample of water in a series of non-reacting glass bottles and analyzing the contents of each bottle at different points in time. If the reaction is first-order, then plotting the natural log (C_t/Co) against time should result in a straight line, where C_t is concentration at time t and Co is concentration at time zero. K_b would then be estimated as the slope of this line.

Bulk reaction coefficients usually increase with increasing temperature. Running multiple bottle tests at different temperatures will provide more accurate assessment of how the rate coefficient varies with temperature

Wall Reactions

The rate of water quality reactions occurring at or near the pipe wall can be considered to be dependent on the concentration in the bulk flow by using an expression of the form

$$R = (A/V) K_w C^n$$

where K_w = a wall reaction rate coefficient and (A/V) = the surface area per unit volume within a pipe (equal to 4 divided by the pipe diameter). The latter term converts the mass reacting per unit of wall area to a per unit volume basis. EPANET limits the choice of wall reaction order to either 0 or 1, so that the units of K_w are either mass/area/time or length/time, respectively. As with K_b, K_w must be supplied to the program by the modeler. First-order K_w values can range anywhere from 0 to as much as 5 ft/day.

K_w should be adjusted to account for any mass transfer limitations in moving reactants and products between the bulk flow and the wall. EPANET does this automatically, basing the adjustment on the molecular diffusivity of the substance being modeled and on the flow's Reynolds number. See Appendix D for details. (Setting the molecular diffusivity to zero will cause mass transfer effects to be ignored.)

The wall reaction coefficient can depend on temperature and can also be correlated to pipe age and material. It is well known that as metal pipes age their roughness tends to increase due to encrustation and tuburculation of corrosion products on the pipe walls. This increase in roughness produces a lower Hazen-Williams C-factor or a higher Darcy-Weisbach roughness coefficient, resulting in greater frictional head loss in flow through the pipe.

There is some evidence to suggest that the same processes that increase a pipe's roughness with age also tend to increase the reactivity of its wall with some chemical species, particularly chlorine and other disinfectants. EPANET can make each pipe's K_w be a function of the coefficient used to describe its roughness. A different function applies depending on the formula used to compute headloss through the pipe:

Headloss Formula	*Wall Reaction Formula*
Hazen-Williams	$K_w = F / C$
Darcy-Weisbach	$K_w = -F / \log(e/d)$
Chezy-Manning	$K_w = F\, n$

where C = Hazen-Williams C-factor, e = Darcy-Weisbach roughness, d = pipe diameter, n = Manning roughness coefficient, and F = wall reaction - pipe roughness coefficient The coefficient F must be developed from site-specific field measurements and will have a different meaning depending on which head loss equation is used. The advantage of using this approach is that it requires only a single parameter, F, to allow wall reaction coefficients to vary throughout the network in a physically meaningful way.

Water Age and Source Tracing

In addition to chemical transport, EPANET can also model the changes in the age of water throughout a distribution system. Water age is the time spent by a parcel of water in the network. New water entering the network from reservoirs or source nodes enters with age of zero. Water age provides a simple, non-specific measure of the overall quality of delivered drinking water. Internally, EPANET treats age as a reactive constituent whose growth follows zero-order kinetics with a rate constant equal to 1 (i.e., each second the water becomes a second older).

EPANET can also perform source tracing. Source tracing tracks over time what percent of water reaching any node in the network had its origin at a particular node. The source node can be any node in the network, including tanks or reservoirs. Internally, EPANET treats this node as a constant source of a non-reacting constituent that enters the network with a concentration of 100. Source tracing is a useful tool for analyzing distribution systems drawing water from two or more different raw water supplies. It can show to what degree water from a given source blends with that from other sources, and how the spatial pattern of this blending changes over time.

CHAPTER 4 - EPANET'S WORKSPACE

This chapter discusses the essential features of EPANET's workspace. It describes the main menu bar, the tool and status bars, and the three windows used most often – the Network Map, the Browser, and the Property Editor. It also shows how to set program preferences.

4.1 Overview

The basic EPANET workspace is pictured below. It consists of the following user interface elements: a Menu Bar, two Toolbars, a Status Bar, the Network Map window, a Browser window, and a Property Editor window. A description of each of these elements is provided in the sections that follow.

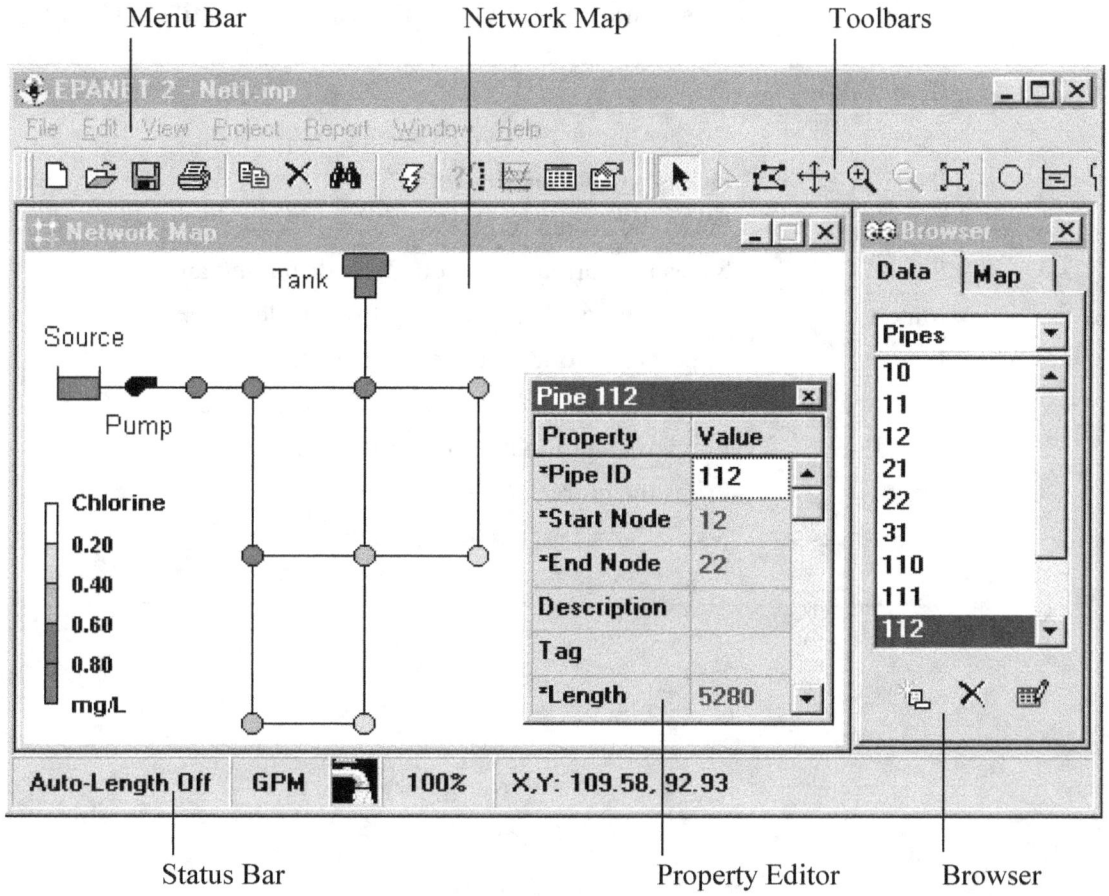

4.2 Menu Bar

The Menu Bar located across the top of the EPANET workspace contains a collection of menus used to control the program. These include:

- File Menu
- Edit Menu
- View Menu
- Project Menu
- Report Menu
- Window Menu
- Help Menu

File Menu

The File Menu contains commands for opening and saving data files and for printing:

Command	Description
New	Creates a new EPANET project
Open	Opens an existing project
Save	Saves the current project
Save As	Saves the current project under a different name
Import	Imports network data or map from a file
Export	Exports network data or map to a file
Page Setup	Sets page margins, headers, and footers for printing
Print Preview	Previews a printout of the current view
Print	Prints the current view
Preferences	Sets program preferences
Exit	Exits EPANET

Edit Menu

The Edit Menu contains commands for editing and copying.

Command	Description
Copy To	Copies the currently active view (map, report, graph or table) to the clipboard or to file
Select Object	Allows selection of an object on the map
Select Vertex	Allows selection of link vertices on the map
Select Region	Allows selection of an outlined region on the map
Select All	Makes the outlined region the entire viewable map area
Group Edit	Edits a property for the group of objects that fall within the outlined region of the map

View Menu

The View Menu controls how the network map is viewed.

Command	Description
Dimensions	Dimensions the map
Backdrop	Allows a backdrop map to be viewed
Pan	Pans across the map
Zoom In	Zooms in on the map
Zoom Out	Zooms out on the map
Full Extent	Redraws the map at full extent
Find	Locates a specific item on the map
Query	Searches for items on the map that meet specific criteria
Overview Map	Toggles the Overview Map on/off
Legends	Controls the display of map legends
Toolbars	Toggles the toolbars on/off
Options	Sets map appearance options

Project Menu

The Project menu includes commands related to the current project being analyzed.

Command	Description
Summary	Provides a summary description of the project's characteristics
Defaults	Edits a project's default properties
Calibration Data	Registers files containing calibration data with the project
Analysis Options	Edits analysis options
Run Analysis	Runs a simulation

Report Menu

The Report menu has commands used to report analysis results in different formats.

Command	Description
Status	Reports changes in the status of links over time
Energy	Reports the energy consumed by each pump
Calibration	Reports differences between simulated and measured values
Reaction	Reports average reaction rates throughout the network
Full	Creates a full report of computed results for all nodes and links in all time periods which is saved to a plain text file
Graph	Creates time series, profile, frequency, and contour plots of selected parameters
Table	Creates a tabular display of selected node and link quantities
Options	Controls the display style of a report, graph, or table

Window Menu

The Window Menu contains the following commands:

Command	Description
Arrange	Rearranges all child windows to fit within the main window
Close All	Closes all open windows (except the Map and Browser)
Window List	Lists all open windows; selected window currently has focus

Help Menu

The Help Menu contains commands for getting help in using EPANET:

Command	Description
Help Topics	Displays the Help system's Help Topics dialog box
Units	Lists the units of measurement for all EPANET parameters
Tutorial	Presents a short tutorial introducing the user to EPANET
About	Lists information about the version of EPANET being used

Context-sensitive Help is also available by pressing the F1 key.

4.3 Toolbars

Toolbars provide shortcuts to commonly used operations. There are two such toolbars:

- Standard Toolbar
- Map Toolbar

The toolbars can be docked underneath the Main Menu bar or dragged to any location on the EPANET workspace. When undocked, they can also be re-sized. The toolbars can be made visible or invisible by selecting **View >> Toolbars**.

Standard Toolbar

The Standard Toolbar contains speed buttons for commonly used commands.

Opens a new project (**File >> New**)

Opens an existing project (**File >> Open**)

Saves the current project (**File >> Save**)

Prints the currently active window (**File >> Print**)

Copies selection to the clipboard or to a file (**Edit >> Copy To**)

Deletes the currently selected item

Finds a specific item on the map (**View >> Find**)

Runs a simulation (**Project >> Run Analysis**)

Runs a visual query on the map (**View >> Query**)

Creates a new graph view of results (**Report >> Graph**)

Creates a new table view of results (**Report >> Table**)

Modifies options for the currently active view (**View >> Options** or **Report >> Options**)

Map Toolbar

The Map Toolbar contains buttons for working with the Network Map.

Selects an object on the map (**Edit >> Select Object**)

Selects link vertex points (**Edit >> Select Vertex**)

Selects a region on the map (**Edit >> Select Region**)

Pans across the map (**View >> Pan**)

Zooms in on the map (**View >> Zoom In**)

Zooms out on the map (**View >> Zoom Out**)

Draws map at full extent (**View >> Full Extent**)

Adds a junction to the map

Adds a reservoir to the map

Adds a tank to the map

Adds a pipe to the map

Adds a pump to the map

Adds a valve to the map

Adds a label to the map

4.4 Status Bar

The Status Bar appears at the bottom of the EPANET workspace and is divided into four sections which display the following information:

- **Auto-Length** – indicates whether automatic computation of pipe lengths is turned on or off

- **Flow Units** - displays the current flow units that are in effect

- **Zoom Level** - displays the current zoom in level for the map (100% is full scale)

- **Run Status** - a faucet icon shows:

    - no running water if no analysis results are available,

- running water when valid analysis results are available,
- a broken faucet when analysis results are available but may be invalid because the network data have been modified.

- **XY Location** - displays the map coordinates of the current position of the mouse pointer.

4.5 Network Map

The Network Map provides a planar schematic diagram of the objects comprising a water distribution network. The location of objects and the distances between them do not necessarily have to conform to their actual physical scale. Selected properties of these objects, such as water quality at nodes or flow velocity in links, can be displayed by using different colors. The color-coding is described in a Legend, which can be edited. New objects can be directly added to the map and existing objects can be clicked on for editing, deleting, and repositioning. A backdrop drawing (such as a street or topographic map) can be placed behind the network map for reference. The map can be zoomed to any scale and panned from one position to another. Nodes and links can be drawn at different sizes, flow direction arrows added, and object symbols, ID labels and numerical property values displayed. The map can be printed, copied onto the Windows clipboard, or exported as a DXF file or Windows metafile.

4.6 Data Browser

The Data Browser (shown below) is accessed from the Data tab on the Browser window. It gives access to the various objects, by category (Junctions, Pipes, etc.) that are contained in the network being analyzed. The buttons at the bottom are used to add, delete, and edit these objects.

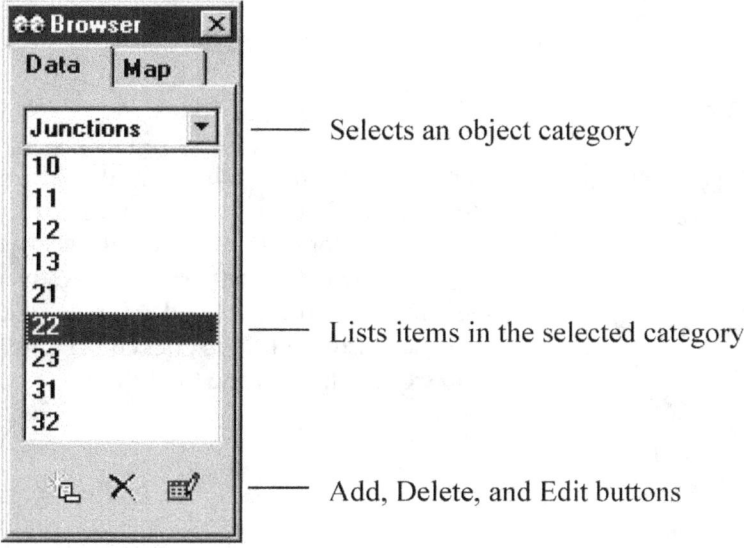

4.7 Map Browser

The Map Browser (shown below) is accessed from the Map tab of the Browser Window. It selects the parameters and time period that are viewed in color-coded fashion on the Network Map. It also contains controls for animating the map through time.

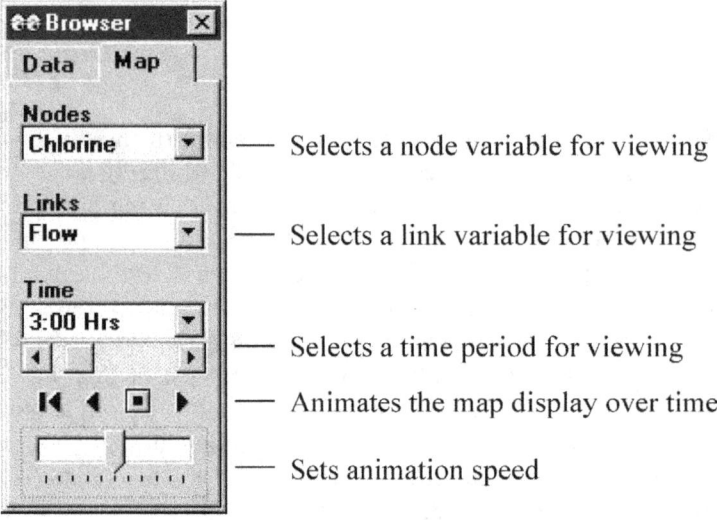

The animation control pushbuttons on the Map Browser work as follows:

⏮	Rewind (return to initial time)
◀	Animate back through time
◼	Stop the animation
▶	Animate forward in time

4.8 Property Editor

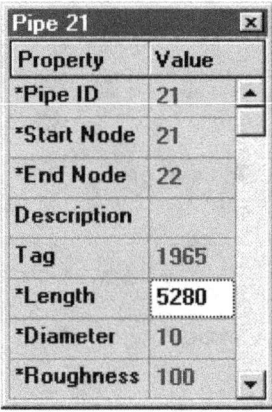

The Property Editor (shown at the left) is used to edit the properties of network nodes, links, labels, and analysis options. It is invoked when one of these objects is selected (either on the Network Map or in the Data Browser) and double-clicked or the Browser's Edit button is clicked. The following points help explain how to use the Editor.

- The Editor is a grid with two columns - one for the property's name and the other for its value.
- The columns can be re-sized by re-sizing the header at the top of the Editor with the mouse.
- The Editor window can be moved and re-sized via the normal Windows procedures.
- An asterisk next to a property name means that it is a required property -- its value cannot be left blank.
- Depending on the property, the value field can be one of the following:
    - a text box where you type in a value
    - a dropdown list box where you select from a list of choices
    - an ellipsis button which you click to bring up a specialized editor
    - a read-only label used to display computed results
- The property in the Editor that currently has focus will be highlighted with a white background.
- You can use both the mouse and the Up and Down arrow keys on the keyboard to move between properties.
- To begin editing the property with the focus, either begin typing a value or hit the Enter key.
- To have EPANET accept what you have entered, press the Enter key or move to another property; to cancel, press the Esc key.
- Clicking the Close button in the upper right corner of its title bar will hide the Editor.

4.9 Program Preferences

Program preferences allow you to customize certain program features. To set program preferences select **Preferences** from the **File** menu. A Preferences dialog form will appear containing two tabbed pages – one for General Preferences and one for Format Preferences.

General Preferences

The following preferences can be set on the General page of the Preferences dialog:

Preference	Description
Bold Fonts	Check to use bold fonts in all newly created windows
Blinking Map Hiliter	Check to make the selected node, link, or label on the map blink on and off
Flyover Map Labeling	Check to display the ID label and current parameter value in a hint-style box whenever the mouse is placed over a node or link on the network map
Confirm Deletions	Check to display a confirmation dialog box before deleting any object
Automatic Backup File	Check to save a backup copy of a newly opened project to disk named with a .bak extension
Temporary Directory	Name of the directory (folder) where EPANET writes its temporary files

Note: The Temporary Directory must be a file directory (folder) where the user has write privileges and must have sufficient space to store files which can easily grow to several tens of megabytes for larger networks and simulation runs. The original default is the Windows TEMP directory (usually c:\Windows\Temp).

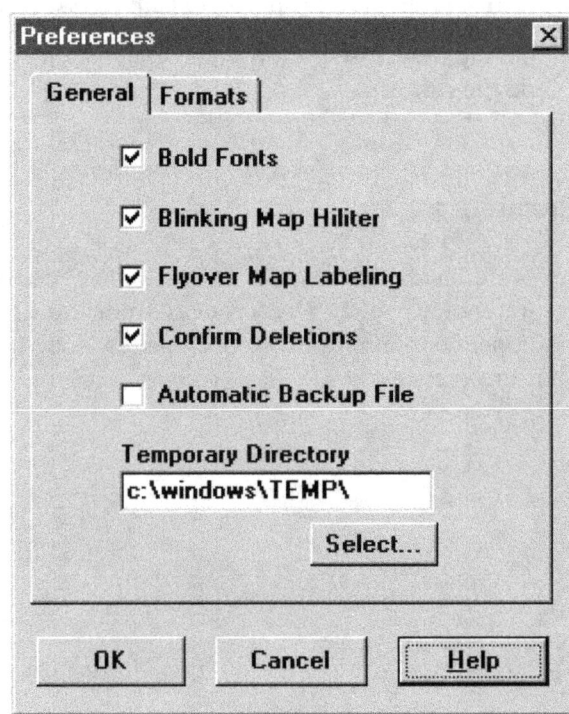

Formatting Preferences

The Formats page of the Preferences dialog box controls how many decimal places are displayed when results for computed parameters are reported. Use the dropdown list boxes to select a specific Node or Link parameter. Use the spin edit boxes to select the number of decimal places to use when displaying computed results for the parameter. The number of decimal places displayed for any particular input design parameter, such as pipe diameter, length, etc. is whatever the user enters.

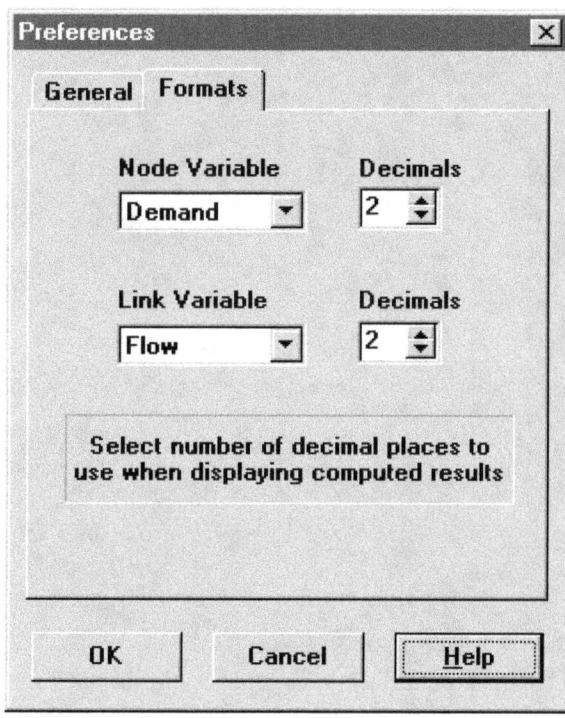

(This page intentionally left blank.)

CHAPTER 5 - WORKING WITH PROJECTS

This chapter discusses how EPANET uses project files to store a piping network's data. It explains how to set certain default options for the project and how to register calibration data (observed measurements) with the project to use for model evaluation.

5.1 Opening and Saving Project Files

Project files contain all of the information used to model a network. They are usually named with a .NET extension.

To create a new project:

1. Select **File >> New** from the Menu Bar or click on the Standard Toolbar.
2. You will be prompted to save the existing project (if changes were made to it) before the new project is created.
3. A new, unnamed project is created with all options set to their default values.

A new project is automatically created whenever EPANET first begins.

To open an existing project stored on disk:

1. Either select **File >> Open** from the Menu Bar or click on the Standard Toolbar.
2. You will be prompted to save the current project (if changes were made to it).
3. Select the file to open from the Open File dialog form that will appear. You can choose to open a file type saved previously as an EPANET project (typically with a .NET extension) or exported as a text file (typically with a .INP extension). EPANET recognizes file types by their content, not their names.
4. Click **OK** to close the dialog and open the selected file.

To save a project under its current name:

- Either select **File >> Save** from the Menu Bar or click on the Standard Toolbar.

To save a project using a different name:

1. Select **File >> Save As** from the Menu Bar.
2. A standard File Save dialog form will appear from which you can select the folder and name that the project should be saved under.

Note: Projects are always saved as binary .NET files. To save a project's data as readable ASCII text, use the **Export >> Network** command from the **File** menu.

5.2 Project Defaults

Each project has a set of default values that are used unless overridden by the EPANET user. These values fall into three categories:

- Default ID labels (labels used to identify nodes and links when they are first created)
- Default node/link properties (e.g., node elevation, pipe length, diameter, and roughness)
- Default hydraulic analysis options (e.g., system of units, headloss equation, etc.)

To set default values for a project:

1. Select **Project >> Defaults** from the Menu Bar.
2. A Defaults dialog form will appear with three pages, one for each category listed above.
3. Check the box in the lower right of the dialog form if you want to save your choices for use in all new future projects as well.
4. Click **OK** to accept your choice of defaults.

The specific items for each category of defaults will be discussed next.

Default ID Labels

The ID Labels page of the Defaults dialog form is shown in Figure 5.1 below. It is used to determine how EPANET will assign default ID labels to network components when they are first created. For each type of object one can enter a label prefix or leave the field blank if the default ID will simply be a number. Then one supplies an increment to be used when adding a numerical suffix to the default label. As an example, if J were used as a prefix for Junctions along with an increment of 5, then as junctions are created they receive default labels of J5, J10, J15 and so on. After an object has been created, the Property Editor can be used to modify its ID label if need be.

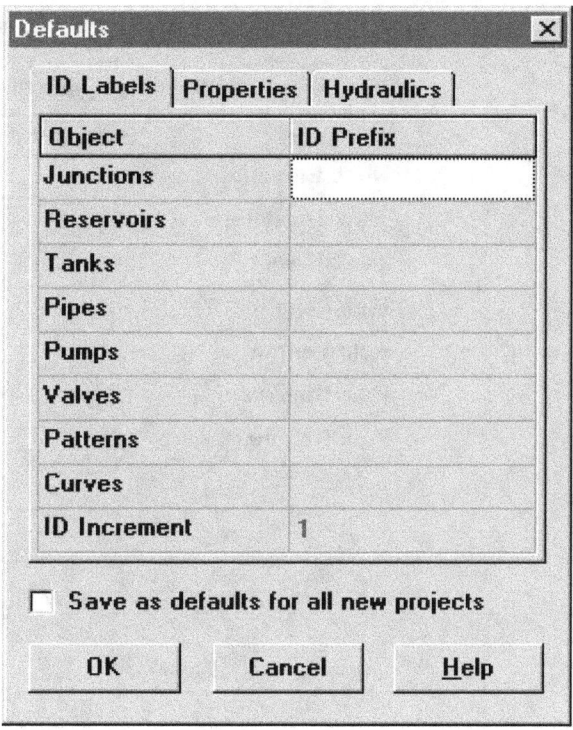

Figure 5.1 ID Labels Page of Project Defaults Dialog

Default Node/Link Properties

The Properties page of the Defaults dialog form is shown in Figure 5.2. It sets default property values for newly created nodes and links. These properties include:

- Elevation for nodes
- Diameter for tanks
- Maximum water level for tanks
- Length for pipes
- Auto-Length (automatic calculation of length) for pipes
- Diameter for pipes
- Roughness for pipes

When the Auto-Length property is turned on, pipe lengths will automatically be computed as pipes are added or repositioned on the network map. A node or link created with these default properties can always be modified later on using the Property Editor.

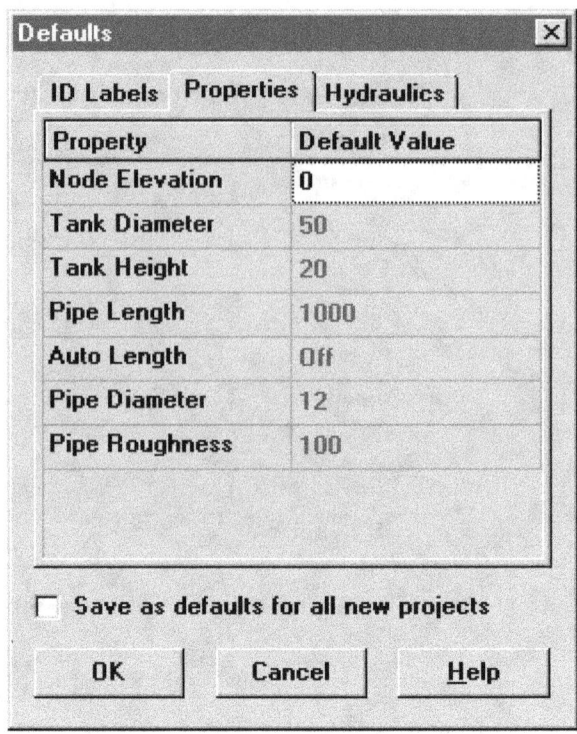

Figure 5.2 Properties Page of the Project Defaults Dialog

Default Hydraulic Options

The third page of the Defaults dialog form is used to assign default hydraulic analysis options. It contains the same set of hydraulic options as the project's Hydraulic Options accessed from the Browser (see Section 8.1). They are repeated on the Project Defaults dialog so that they can be saved for use with future projects as well as with the current one. The most important Hydraulic Options to check when setting up a new project are Flow Units, Headloss Formula, and Default Pattern. The choice of Flow Units determines whether all other network quantities are expressed in Customary US units or in SI metric units. The choice of Headloss Formula defines the type of the roughness coefficient to be supplied for each pipe in the network. The Default Pattern automatically becomes the time pattern used to vary demands in an extended period simulation for all junctions not assigned any pattern.

5.3 Calibration Data

EPANET allows you to compare results of a simulation against measured field data. This can be done via Time Series plots for selected locations in the network or by special Calibration Reports that consider multiple locations. Before EPANET can use such calibration data it has to be entered into a file and registered with the project.

Calibration Files

A Calibration File is a text file containing measured data for a particular quantity taken over a particular period of time within a distribution system. The file provides observed data that can be compared to the results of a network simulation. Separate files should be created for different parameters (e.g., pressure, fluoride, chlorine, flow, etc.) and different sampling studies. Each line of the file contains the following items:

- Location ID - ID label (as used in the network model) of the location where the measurement was made
- Time - Time (in hours) when the measurement was made
- Value - Result of the measurement

The measurement time is with respect to time zero of the simulation to which the Calibration File will be applied. It can be entered as either a decimal number (e.g., 27.5) or in hours:minutes format (e.g., 27:30). For data to be used in a single period analysis all time values can be 0. Comments can be added to the file by placing a semicolon (;) before them. For a series of measurements made at the same location the Location ID does not have to be repeated. An excerpt from a Calibration File is shown below.

```
;Fluoride Tracer Measurements
;Location    Time      Value
;---------------------------
    N1        0         0.5
              6.4       1.2
             12.7       0.9
    N2        0.5       0.72
              5.6       0.77
```

Registering Calibration Data

To register calibration data residing in a Calibration File:

1. Select **Project >> Calibration Data** from the Menu Bar.
2. In the Calibration Data dialog form shown in Figure 5.3, click in the box next to the parameter you wish to register data for.
3. Either type in the name of a Calibration File for this parameter or click the **Browse** button to search for it.
4. Click the **Edit** button if you want to open the Calibration File in Windows NotePad for editing.
5. Repeat steps 2 - 4 for any other parameters that have calibration data.
6. Click **OK** to accept your selections.

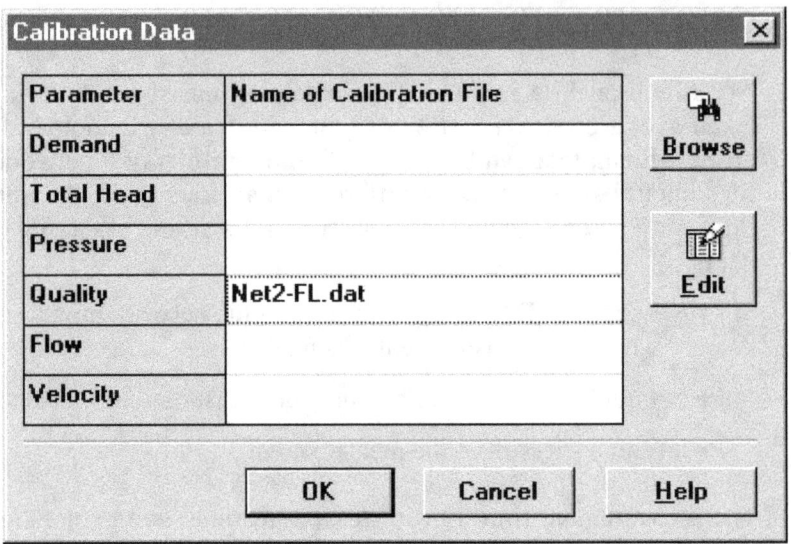

Figure 5.3 Calibration Data Dialog

5.4 Project Summary

To view a summary description of the current project select **Project >> Summary** from the Menu Bar. The Project Summary dialog form will appear in which you can edit a descriptive title for the project as well as add notes that further describe the project. When you go to open a previously saved file, the Open File dialog box will display both of these items as different file names are selected. This makes them very useful for locating specific network analyses. The form also displays certain network statistics, such as the number of junctions, pipes, pumps, etc.

CHAPTER 6 - WORKING WITH OBJECTS

EPANET uses various types of objects to model a distribution system. These objects can be accessed either directly on the network map or from the Data page of the Browser window. This chapter describes what these objects are and how they can be created, selected, edited, deleted, and repositioned.

6.1 Types of Objects

EPANET contains both physical objects that can appear on the network map, and non-physical objects that encompass design and operational information. These objects can be classified as followed:

(1) Nodes
 (a) Junctions
 (b) Reservoirs
 (c) Tanks

(2) Links
 (a) Pipes
 (b) Pumps
 (c) Valves

(3) Map Labels

(4) Time Patterns

(5) Curves

(6) Controls
 (a) Simple
 (b) Rule-Based

6.2 Adding Objects

Adding a Node

To add a Node using the Map Toolbar:

1. Click the button for the type of node (junction , reservoir , or tank) to add from the Map Toolbar if it is not already depressed.
2. Move the mouse to the desired location on the map and click.

To add a Node using the Browser:

1. Select the type of node (junction, reservoir, or tank) from the Object list of the Data Browser.
2. Click the Add button .
3. Enter map coordinates with the Property Editor (optional).

Adding a Link

To add a straight or curved-line Link using the Map Toolbar:

1. Click the button for the type of link to add (pipe , pump , or valve) from the Map Toolbar if it is not already depressed.
2. On the map, click the mouse over the link's start node.
3. Move the mouse in the direction of the link's end node, clicking it at those intermediate points where it is necessary to change the link's direction.
4. Click the mouse a final time over the link's end node.

Pressing the right mouse button or the Escape key while drawing a link will cancel the operation.

To add a straight line Link using the Browser:

1. Select the type of link to add (pipe, pump, or valve) from the Object list of the Data Browser.
2. Click the Add button.
3. Enter the From and To nodes of the link in the Property Editor.

Adding a Map Label

To add a label to the map:

1. Click the Text button on the Map Toolbar.
2. Click the mouse on the map where label should appear.
3. Enter the text for the label.
4. Press the **Enter** key.

Adding a Curve

To add a curve to the network database:

1. Select Curve from the object category list of the Data Browser.
2. Click the Add button.
3. Edit the curve using the Curve Editor (see below).

Adding a Time Pattern

To add a time pattern to the network:

1. Select Patterns from the object category list of the Data Browser.
2. Click the Add button.
3. Edit the pattern using the Pattern Editor (see below).

Using a Text File

In addition to adding individual objects interactively, you can import a text file containing a list of node ID's with their coordinates as well as a list of link ID's and their connecting nodes (see Section 11.4 - Importing a Partial Network).

6.3 Selecting Objects

To select an object on the map:

1. Make sure that the map is in Selection mode (the mouse cursor has the shape of an arrow pointing up to the left). To switch to this mode, either click the Select Object button on the Map Toolbar or choose **Select Object** from the **Edit** menu.
2. Click the mouse over the desired object on the map.

To select an object using the Browser:

1. Select the category of object from the dropdown list of the Data Browser.
2. Select the desired object from the list below the category heading.

6.4 Editing Visual Objects

The Property Editor (see Section 4.8) is used to edit the properties of objects that can appear on the Network Map (Junctions, Reservoirs, Tanks, Pipes, Pumps, Valves, or Labels). To edit one of these objects, select the object on the map or from the Data Browser, then click the Edit button on the Data Browser (or simply double-click the object on the map). The properties associated with each of these types of objects are described in Tables 6.1 to 6.7.

Note: The unit system in which object properties are expressed depends on the choice of units for flow rate. Using a flow rate expressed in cubic feet, gallons or acre-feet means that US units will be used for all quantities. Using a flow rate expressed in liters or cubic meters means that SI metric units will be used. Flow units are selected from the project's Hydraulic Options which can be accessed from the **Project >> Defaults** menu. The units used for all properties are summarized in Appendix A.

Table 6.1 Junction Properties

PROPERTY	DESCRIPTION
Junction ID	A unique label used to identify the junction. It can consist of a combination of up to 15 numerals or characters. It cannot be the same as the ID for any other node. This is a required property.
X-Coordinate	The horizontal location of the junction on the map, measured in the map's distance units. If left blank the junction will not appear on the network map.
Y-Coordinate	The vertical location of the junction on the map, measured in the map's distance units. If left blank the junction will not appear on the network map.
Description	An optional text string that describes other significant information about the junction.
Tag	An optional text string (with no spaces) used to assign the junction to a category, such as a pressure zone.
Elevation	The elevation in feet (meters) above some common reference of the junction. This is a required property. Elevation is used only to compute pressure at the junction. It does not affect any other computed quantity.
Base Demand	The average or nominal demand for water by the main category of consumer at the junction, as measured in the current flow units. A negative value is used to indicate an external source of flow into the junction. If left blank then demand is assumed to be zero.
Demand Pattern	The ID label of the time pattern used to characterize time variation in demand for the main category of consumer at the junction. The pattern provides multipliers that are applied to the Base Demand to determine actual demand in a given time period. If left blank then the **Default Time Pattern** assigned in the Hydraulic Options (see Section 8.1) will be used.
Demand Categories	Number of different categories of water users defined for the junction. Click the ellipsis button (or hit the Enter key) to bring up a special Demands Editor which will let you assign base demands and time patterns to multiple categories of users at the junction. Ignore if only a single demand category will suffice.
Emitter Coefficient	Discharge coefficient for emitter (sprinkler or nozzle) placed at junction. The coefficient represents the flow (in current flow units) that occurs at a pressure drop of 1 psi (or meter). Leave blank if no emitter is present. See the Emitters topic in Section 3.1 for more details.
Initial Quality	Water quality level at the junction at the start of the simulation period. Can be left blank if no water quality analysis is being made or if the level is zero.
Source Quality	Quality of any water entering the network at this location. Click the ellipsis button (or hit the Enter key) to bring up the Source Quality Editor (see Section 6.5 below).

Table 6.2 Reservoir Properties

PROPERTY	DESCRIPTION
Reservoir ID	A unique label used to identify the reservoir. It can consist of a combination of up to 15 numerals or characters. It cannot be the same as the ID for any other node. This is a required property.
X-Coordinate	The horizontal location of the reservoir on the map, measured in the map's distance units. If left blank the reservoir will not appear on the network map.
Y-Coordinate	The vertical location of the reservoir on the map, measured in the map's distance units. If left blank the reservoir will not appear on the network map.
Description	An optional text string that describes other significant information about the reservoir.
Tag	An optional text string (with no spaces) used to assign the reservoir to a category, such as a pressure zone
Total Head	The hydraulic head (elevation + pressure head) of water in the reservoir in feet (meters). This is a required property.
Head Pattern	The ID label of a time pattern used to model time variation in the reservoir's head. Leave blank if none applies. This property is useful if the reservoir represents a tie-in to another system whose pressure varies with time.
Initial Quality	Water quality level at the reservoir. Can be left blank if no water quality analysis is being made or if the level is zero.
Source Quality	Quality of any water entering the network at this location. Click the ellipsis button (or hit the Enter key) to bring up the Source Quality Editor (see Section 6.5 below).

Table 6.3 Tank Properties

PROPERTY	DESCRIPTION
Tank ID	A unique label used to identify the tank. It can consist of a combination of up to 15 numerals or characters. It cannot be the same as the ID for any other node. This is a required property.
X-Coordinate	The horizontal location of the tank on the map, measured in the map's scaling units. If left blank the tank will not appear on the network map.
Y-Coordinate	The vertical location of the tank on the map, measured in the map's scaling units. If left blank the tank will not appear on the network map.
Description	Optional text string that describes other significant information about the tank.
Tag	Optional text string (with no spaces) used to assign the tank to a category, such as a pressure zone
Elevation	Elevation above a common datum in feet (meters) of the bottom shell of the tank. This is a required property.
Initial Level	Height in feet (meters) of the water surface above the bottom elevation of the tank at the start of the simulation. This is a required property.
Minimum Level	Minimum height in feet (meters) of the water surface above the bottom elevation that will be maintained. The tank will not be allowed to drop below this level. This is a required property.

Maximum Level	Maximum height in feet (meters) of the water surface above the bottom elevation that will be maintained. The tank will not be allowed to rise above this level. This is a required property.
Diameter	The diameter of the tank in feet (meters). For cylindrical tanks this is the actual diameter. For square or rectangular tanks it can be an equivalent diameter equal to 1.128 times the square root of the cross-sectional area. For tanks whose geometry will be described by a curve (see below) it can be set to any value. This is a required property.
Minimum Volume	The volume of water in the tank when it is at its minimum level, in cubic feet (cubic meters). This is an optional property, useful mainly for describing the bottom geometry of non-cylindrical tanks where a full volume versus depth curve will not be supplied (see below).
Volume Curve	The ID label of a curve used to describe the relation between tank volume and water level. If no value is supplied then the tank is assumed to be cylindrical.
Mixing Model	The type of water quality mixing that occurs within the tank. The choices include - MIXED (fully mixed), - 2COMP (two-compartment mixing), - FIFO (first-in-first-out plug flow), - LIFO (last-in-first-out plug flow). See the Mixing Models topic in Section 3.4 for more information.
Mixing Fraction	The fraction of the tank's total volume that comprises the inlet-outlet compartment of the two-compartment (2COMP) mixing model. Can be left blank if another type of mixing model is employed.
Reaction Coefficient	The bulk reaction coefficient for chemical reactions in the tank. Time units are 1/days. Use a positive value for growth reactions and a negative value for decay. Leave blank if the Global Bulk reaction coefficient specified in the project's Reactions Options will apply. See Water Quality Reactions in Section 3.4 for more information.
Initial Quality	Water quality level in the tank at the start of the simulation. Can be left blank if no water quality analysis is being made or if the level is zero.
Source Quality	Quality of any water entering the network at this location. Click the ellipsis button (or hit the Enter key) to bring up the Source Quality Editor (see Section 6.5 below).

Table 6.4 Pipe Properties

PROPERTY	DESCRIPTION
Pipe ID	A unique label used to identify the pipe. It can consist of a combination of up to 15 numerals or characters. It cannot be the same as the ID for any other link. This is a required property.
Start Node	The ID of the node where the pipe begins. This is a required property.
End Node	The ID of the node where the pipe ends. This is a required property.
Description	An optional text string that describes other significant information about the pipe.
Tag	An optional text string (with no spaces) used to assign the pipe to a category, perhaps one based on age or material
Length	The actual length of the pipe in feet (meters). This is a required property.
Diameter	The pipe diameter in inches (mm). This is a required property.
Roughness	The roughness coefficient of the pipe. It is unitless for Hazen-Williams or Chezy-Manning roughness and has units of millifeet (mm) for Darcy-Weisbach roughness. This is a required property.
Loss Coefficient	Unitless minor loss coefficient associated with bends, fittings, etc. Assumed 0 if left blank.
Initial Status	Determines whether the pipe is initially open, closed, or contains a check valve. If a check valve is specified then the flow direction in the pipe will always be from the Start node to the End node.
Bulk Coefficient	The bulk reaction coefficient for the pipe. Time units are 1/days. Use a positive value for growth and a negative value for decay. Leave blank if the Global Bulk reaction coefficient from the project's Reaction Options will apply. See Water Quality Reactions in Section 3.4 for more information.
Wall Coefficient	The wall reaction coefficient for the pipe. Time units are 1/days. Use a positive value for growth and a negative value for decay. Leave blank if the Global Wall reaction coefficient from the project's Reactions Options will apply. See Water Quality Reactions in Section 3.4 for more information.

Note: Pipe lengths can be automatically computed as pipes are added or repositioned on the network map if the **Auto-Length** setting is turned on. To toggle this setting On/Off either:

- Select **Project >> Defaults** and edit the Auto-Length field on the Properties page of the Defaults dialog form.

- Right-click over the Auto-Length section of the Status Bar and then click on the popup menu item that appears.

Be sure to provide meaningful dimensions for the network map before using the Auto-Length feature (see Section 7.2).

Table 6.5 Pump Properties

PROPERTY	DESCRIPTION
Pump ID	A unique label used to identify the pump. It can consist of a combination of up to 15 numerals or characters. It cannot be the same as the ID for any other link. This is a required property.
Start Node	The ID of the node on the suction side of the pump. This is a required property
End Node	The ID of the node on the discharge side of the pump. This is a required property
Description	An optional text string that describes other significant information about the pump.
Tag	An optional text string (with no spaces) used to assign the pump to a category, perhaps based on age, size or location
Pump Curve	The ID label of the pump curve used to describe the relationship between the head delivered by the pump and the flow through the pump. Leave blank if the pump will be a constant energy pump (see below).
Power	The power supplied by the pump in horsepower (kw). Assumes that the pump supplies the same amount of energy no matter what the flow is. Leave blank if a pump curve will be used instead. Use when pump curve information is not available.
Speed	The relative speed setting of the pump (unitless). For example, a speed setting of 1.2 implies that the rotational speed of the pump is 20% higher than the normal setting.
Pattern	The ID label of a time pattern used to control the pump's operation. The multipliers of the pattern are equivalent to speed settings. A multiplier of zero implies that the pump will be shut off during the corresponding time period. Leave blank if not applicable.
Initial Status	State of the pump (open or closed) at the start of the simulation period.
Efficiency Curve	The ID label of the curve that represents the pump's wire-to-water efficiency (in percent) as a function of flow rate. This information is used only to compute energy usage. Leave blank if not applicable or if the global pump efficiency supplied with the project's Energy Options (see Section 8.1) will be used.
Energy Price	The average or nominal price of energy in monetary units per kw-hr. Used only for computing the cost of energy usage. Leave blank if not applicable or if the global value supplied with the project's Energy Options (Section 8.1) will be used.
Price Pattern	The ID label of the time pattern used to describe the variation in energy price throughout the day. Each multiplier in the pattern is applied to the pump's Energy Price to determine a time-of-day pricing for the corresponding period. Leave blank if not applicable or if the global pricing pattern specified in the project's Energy Options (Section 8.1) will be used.

Table 6.6 Valve Properties

PROPERTY	DESCRIPTION
ID Label	A unique label used to identify the valve. It can consist of a combination of up to 15 numerals or characters. It cannot be the same as the ID for any other link. This is a required property.
Start Node	The ID of the node on the nominal upstream or inflow side of the valve. (PRVs and PSVs maintain flow in only a single direction.) This is a required property.
End Node	The ID of the node on the nominal downstream or discharge side of the valve. This is a required property.
Description	An optional text string that describes other significant information about the valve.
Tag	An optional text string (with no spaces) used to assign the valve to a category, perhaps based on type or location.
Diameter	The valve diameter in inches (mm). This is a required property.
Type	The valve type (PRV, PSV, PBV, FCV, TCV, or GPV). See Valves in Section 6.1 for descriptions of the various types of valves. This is a required property.
Setting	A required parameter that describes the valve's operational setting. Valve Type Setting Parameter PRV Pressure (psi or m) PSV Pressure (psi or m) PBV Pressure (psi or m) FCV Flow (flow units) TCV Loss Coefficient (unitless) GPV ID of head loss curve
Loss Coefficient	Unitless minor loss coefficient that applies when the valve is completely opened. Assumed 0 if left blank.
Fixed Status	Valve status at the start of the simulation. If set to OPEN or CLOSED then the control setting of the valve is ignored and the valve behaves as an open or closed link, respectively. If set to NONE, then the valve will behave as intended. A valve's fixed status and its setting can be made to vary throughout a simulation by the use of control statements. If a valve's status was fixed to OPEN/CLOSED, then it can be made active again using a control that assigns a new numerical setting to it.

Table 6.7 Map Label Properties

PROPERTY	DESCRIPTION
Text	The label's text.
X-Coordinate	The horizontal location of the upper left corner of the label on the map, measured in the map's scaling units. This is a required property.
Y-Coordinate	The vertical location of the upper left corner of the label on the map, measured in the map's scaling units. This is a required property.
Anchor Node	ID of node that serves as the label's anchor point (see Note 1 below). Leave blank if label will not be anchored.
Meter Type	Type of object being metered by the label (see Note 2 below). Choices are None, Node, or Link.
Meter ID	ID of the object (Node or Link) being metered.
Font	Launches a Font dialog that allows selection of the label's font, size, and style.

Notes:

1. A label's anchor node property is used to anchor the label relative to a given location on the map. When the map is zoomed in, the label will appear the same distance from its anchor node as it did under the full extent view. This feature prevents labels from wandering too far away from the objects they were meant to describe when a map is zoomed.

2. The Meter Type and ID properties determine if the label will act as a meter. Meter labels display the value of the current viewing parameter (chosen from the Map Browser) underneath the label text. The Meter Type and ID must refer to an existing node or link in the network. Otherwise, only the label text appears.

6.5 Editing Non-Visual Objects

Curves, Time Patterns, and Controls have special editors that are used to define their properties. To edit one of these objects, select the object from the Data Browser and then click the Edit button. In addition, the Property Editor for Junctions contains an ellipsis button in the field for Demand Categories that brings up a special Demand Editor when clicked. Similarly, the Source Quality field in the Property Editor for Junctions, Reservoirs, and Tanks has a button that launches a special Source Quality editor. Each of these specialized editors is described next.

Curve Editor

The Curve Editor is a dialog form as shown in Figure 6.1. To use the Curve Editor, enter values for the following items:

Item	Description
Curve ID	ID label of the curve (maximum of 15 numerals or characters)
Description	Optional description of what the curve represents
Curve Type	Type of curve
X-Y Data	X-Y data points for the curve

As you move between cells in the X-Y data table (or press the Enter key) the curve is redrawn in the preview window. For single- and three-point pump curves, the equation generated for the curve will be displayed in the Equation box. Click the **OK** button to accept the curve or the **Cancel** button to cancel your entries. You can also click the **Load** button to load in curve data that was previously saved to file or click the **Save** button to save the current curve's data to a file.

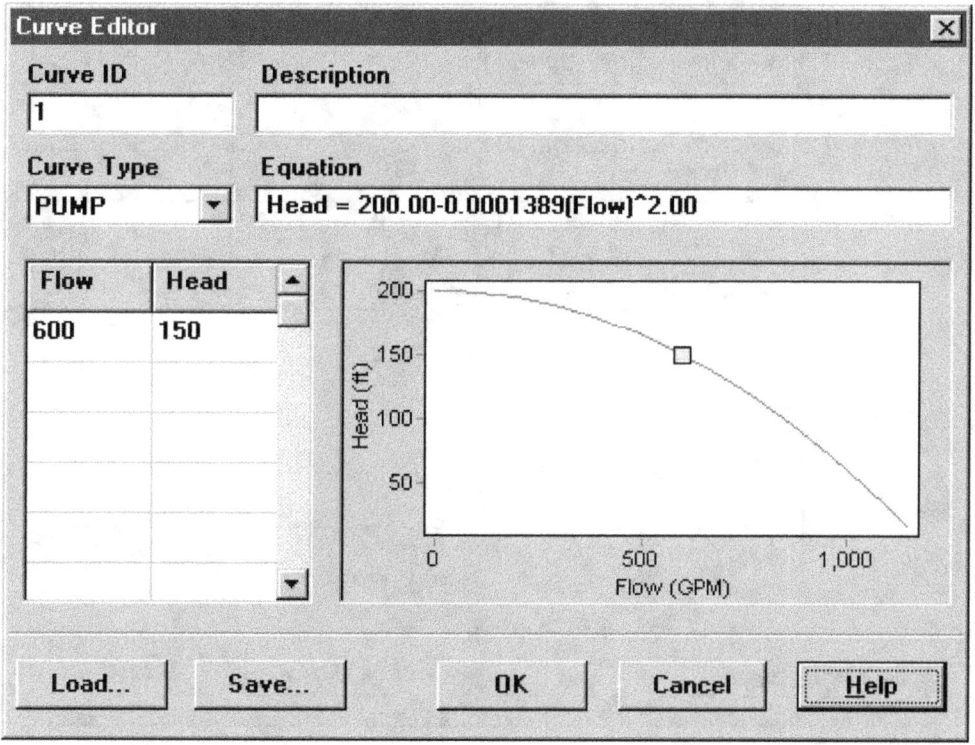

Figure 6.1 Curve Editor

Pattern Editor

The Pattern Editor, displayed in Figure 6.2, edits the properties of a time pattern object. To use the Pattern Editor enter values for the following items:

Item	Description
Pattern ID	ID label of the pattern (maximum of 15 numerals or characters)
Description	Optional description of what the pattern represents
Multipliers	Multiplier value for each time period of the pattern.

As multipliers are entered, the preview chart is redrawn to provide a visual depiction of the pattern. If you reach the end of the available Time Periods when entering multipliers, simply hit the **Enter** key to add on another period. When finished editing, click the **OK** button to accept the pattern or the **Cancel** button to cancel your entries. You can also click the **Load** button to load in pattern data that was previously saved to file or click the **Save** button to save the current pattern's data to a file.

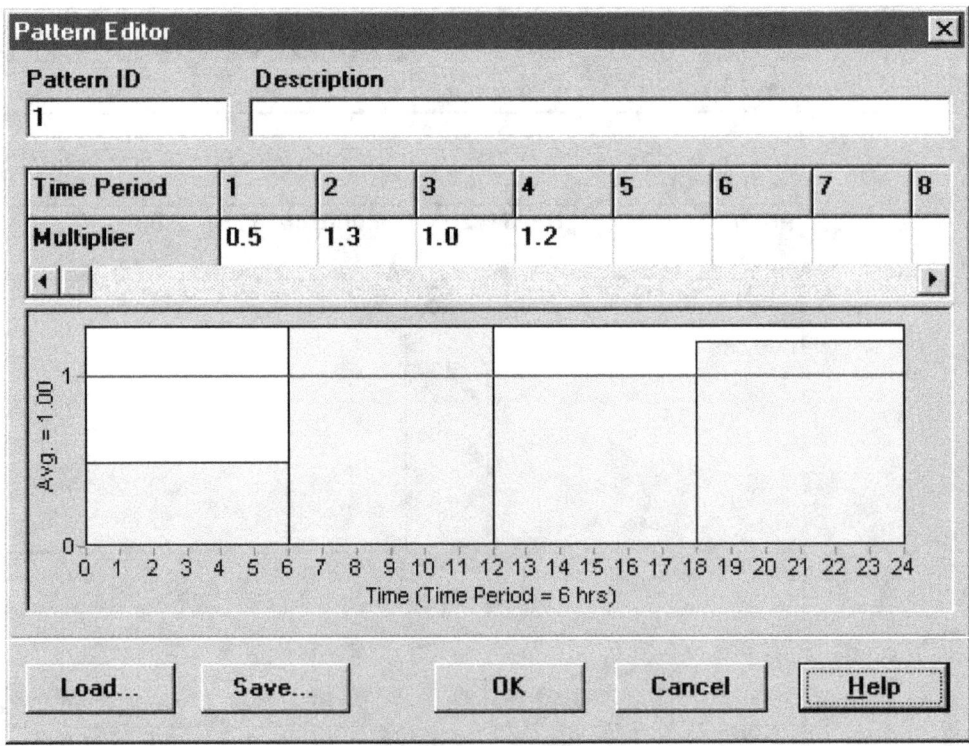

Figure 6.2 Pattern Editor

Controls Editor

The Controls Editor, shown in Figure 6.3, is a text editor window used to edit both simple and rule-based controls. It has a standard text-editing menu that is activated by right-clicking anywhere in the Editor. The menu contains commands for Undo, Cut, Copy, Paste, Delete, and Select All.

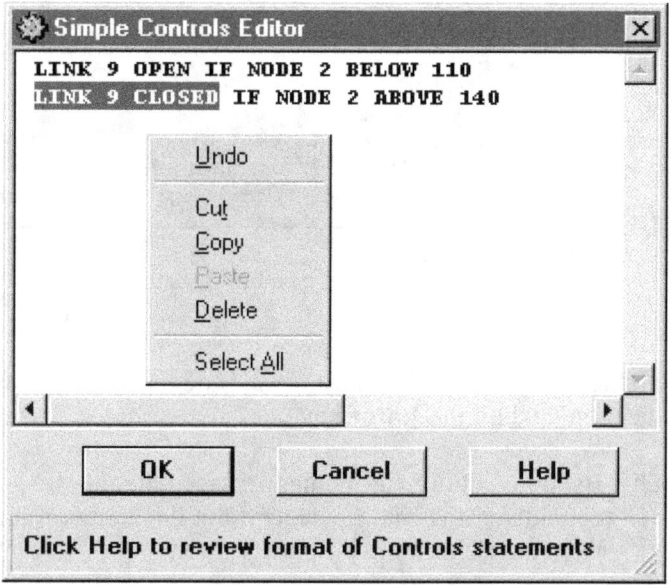

Figure 6.3 Controls Editor

Demand Editor

The Demand Editor is pictured in Figure 6.4. It is used to assign base demands and time patterns when there is more than one category of water user at a junction. The editor is invoked from the Property Editor by clicking the ellipsis button (or hitting the Enter key) when the Demand Categories field has the focus.

The editor is a table containing three columns. Each category of demand is entered as a new row in the table. The columns contain the following information:

- *Base Demand*: baseline or average demand for the category (required)
- *Time Pattern*: ID label of time pattern used to allow demand to vary with time (optional)
- *Category*: text label used to identify the demand category (optional)

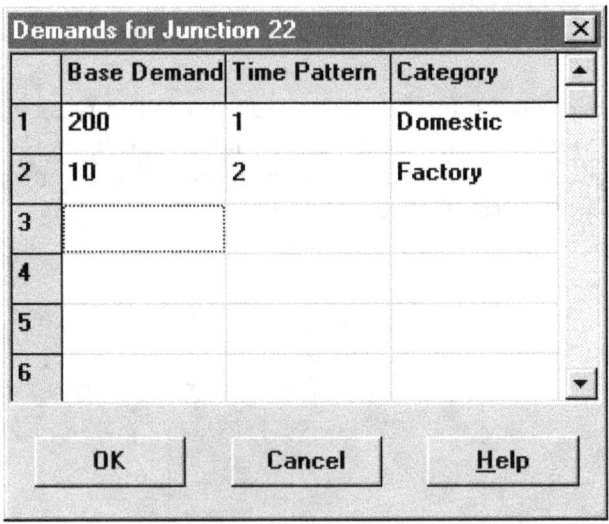

Figure 6.4 Demand Editor

The table initially is sized for 10 rows. If additional rows are needed select any cell in the last row and hit the **Enter** key.

Note: By convention, the demand placed in the first row of the editor will be considered the main category for the junction and will appear in the Base Demand field of the Property Editor.

Source Quality Editor

The Source Quality Editor is a pop-up dialog used to describe the quality of source flow entering the network at a specific node. This source might represent the main treatment works, a well head or satellite treatment facility, or an unwanted contaminant intrusion. The dialog form, shown in Figure 6.5, contains the following fields:

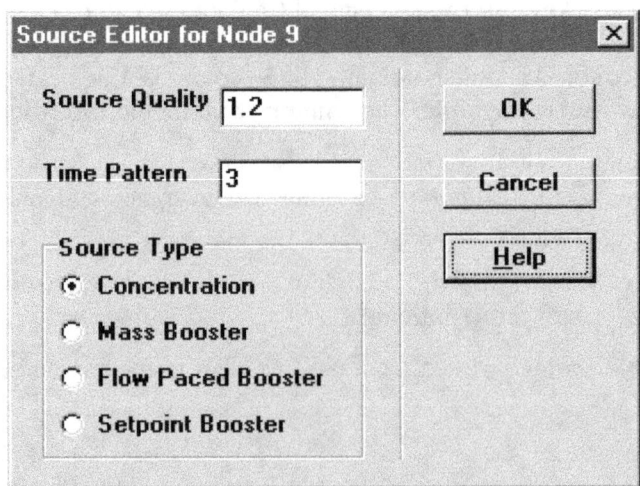

Figure 6.5 Source Quality Editor

Field	Description
Source Type	Select either: - Concentration - Mass Booster - Flow Paced Booster - Setpoint Booster
Source Quality	Baseline or average concentration (or mass flow rate per minute) of source – leave blank to remove the source
Quality Pattern	ID label of time pattern used to make source quality vary with time – leave blank if not applicable

A water quality source can be designated as a concentration or booster source.

- A **concentration source** fixes the concentration of any external inflow entering the network, such as flow from a reservoir or from a negative demand placed at a junction.

- A **mass booster source** adds a fixed mass flow to that entering the node from other points in the network.

- A **flow paced booster source** adds a fixed concentration to that resulting from the mixing of all inflow to the node from other points in the network.

- A **setpoint booster source** fixes the concentration of any flow leaving the node (as long as the concentration resulting from all inflow to the node is below the setpoint).

The concentration-type source is best used for nodes that represent source water supplies or treatment works (e.g., reservoirs or nodes assigned a negative demand). The booster-type source is best used to model direct injection of a tracer or additional disinfectant into the network or to model a contaminant intrusion.

6.6 Copying and Pasting Objects

The properties of an object displayed on the Network Map can be copied and pasted into another object from the same category. To copy the properties of an object to EPANET's internal clipboard:

1. Right-click the object on the map.
2. Select **Copy** from the pop-up menu that appears.

To paste copied properties into an object:

1. Right-click the object on the map.
2. Select **Paste** from the pop-up menu that appears.

6.7 Shaping and Reversing Links

Links can be drawn as polylines containing any number of straight-line segments that add change of direction and curvature to the link. Once a link has been drawn on the map, interior points that define these line segments can be added, deleted, and moved (see Figure 6.6). To edit the interior points of a link:

1. Select the link to edit on the Network Map and click ▷ on the Map Toolbar (or select **Edit >> Select Vertex** from the Menu Bar, or right-click on the link and select **Vertices** from the popup menu).

2. The mouse pointer will change shape to an arrow tip, and any existing vertex points on the link will be displayed with small handles around them. To select a particular vertex, click the mouse over it.

3. To add a new vertex to the link, right-click the mouse and select **Add Vertex** from the popup menu (or simply press the **Insert** key on the keyboard).

4. To delete the currently selected vertex, right-click the mouse and select **Delete Vertex** from the popup menu (or simply press the **Delete** key on the keyboard).

5. To move a vertex to another location, drag it with the left mouse button held down to its new position.

6. While in Vertex Selection mode you can begin editing the vertices for another link by clicking on the link. To leave Vertex Selection mode, right-click on the map and select **Quit Editing** from the popup menu, or select any other button on the Map Toolbar.

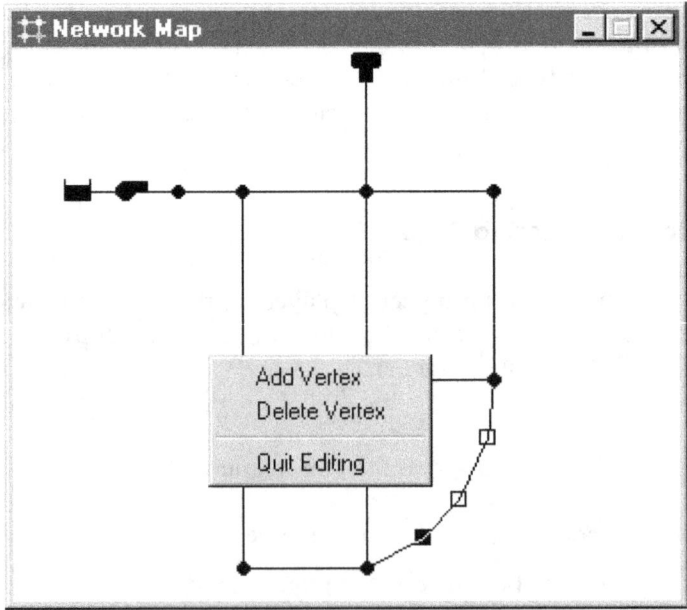

Figure 6.6 Reshaping a Link

A link can also have its direction reversed (i.e., its end nodes switched) by right-clicking on it and selecting **Reverse** from the pop-up menu that appears. This is useful for re-orienting pumps and valves that originally were added in the wrong direction.

6.8 Deleting an Object

To delete an object:
1. Select the object on the map or from the Data Browser.
2. Either:
   - click ☒ on the Standard Toolbar,
   - click the same button on the Data Browser,
   - press the **Delete** key on the keyboard.

Note: You can require that all deletions be confirmed before they take effect. See the General Preferences page of the Program Preferences dialog box described in Section 4.9.

6.9 Moving an Object

To move a node or label to another location on the map:
1. Select the node or label.
2. With the left mouse button held down over the object, drag it to its new location.
3. Release the left button.

Alternatively, new X and Y coordinates for the object can be typed in manually in the Property Editor. Whenever a node is moved all links connected to it are moved as well.

6.10 Selecting a Group of Objects

To select a group of objects that lie within an irregular region of the network map:
1. Select **Edit >> Select Region** or click 🔲 on the Map Toolbar.
2. Draw a polygon fence line around the region of interest on the map by clicking the left mouse button at each successive vertex of the polygon.
3. Close the polygon by clicking the right button or by pressing the **Enter** key; Cancel the selection by pressing the **Escape** key.

To select all objects currently in view on the map select **Edit >> Select All**. (Objects outside the current viewing extent of the map are not selected.)

Once a group of objects has been selected, you can edit a common property (see the following section) or delete the selected objects from the network. To do the latter, click ☒ or press the **Delete** key.

6.11 Editing a Group of Objects

To edit a property for a group of objects:

1. Select the region of the map that will contain the group of objects to be edited using the method described in previous section.
2. Select **Edit >> Group Edit** from the Menu Bar.
3. Define what to edit in the Group Edit dialog form that appears.

The Group Edit dialog form, shown in Figure 6.6, is used to modify a property for a selected group of objects. To use the dialog form:

1. Select a category of object (Junctions or Pipes) to edit.
2. Check the "with" box if you want to add a filter that will limit the objects selected for editing. Select a property, relation and value that define the filter. An example might be "with Diameter below 12".
3. Select the type of change to make - Replace, Multiply, or Add To.
4. Select the property to change.
5. Enter the value that should replace, multiply, or be added to the existing value.
6. Click **OK** to execute the group edit.

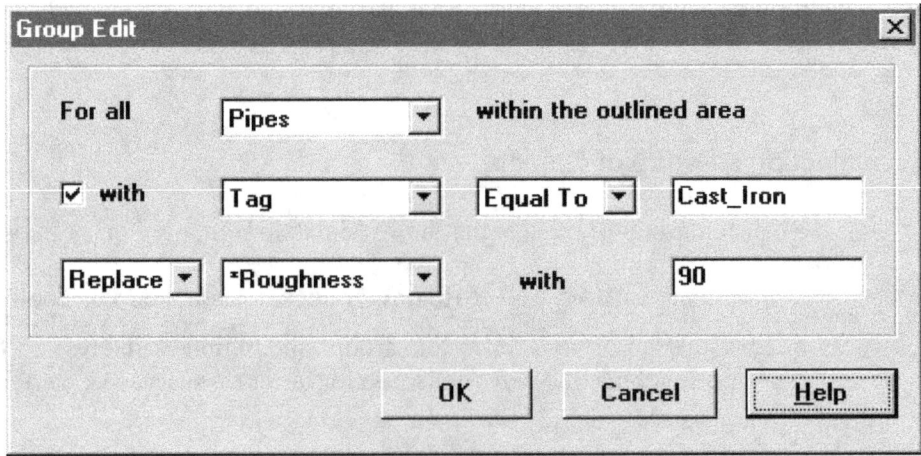

Figure 6.7 Group Edit Dialog

CHAPTER 7 - WORKING WITH THE MAP

EPANET displays a map of the pipe network being modeled. This chapter describes how you can manipulate this map to enhance your visualization of the system being modeled.

7.1 Selecting a Map View

One uses the Map Page of the Browser (Section 4.7) to select a node and link parameter to view on the map. Parameters are viewed on the map by using colors, as specified in the Map Legends (see below), to display different ranges of values.

Node parameters available for viewing include:

- Elevation
- Base Demand (nominal or average demand)
- Initial Quality (water quality at time zero)
- *Actual Demand (total demand at current time)
- *Hydraulic Head (elevation plus pressure head)
- *Pressure
- *Water Quality

Link parameters available for viewing include:

- Length
- Diameter
- Roughness Coefficient
- Bulk Reaction Coefficient
- Wall Reaction Coefficient
- *Flow Rate
- *Velocity
- *Headloss (per 1000 feet (or meters) of pipe)
- *Friction Factor (as used in the Darcy-Weisbach headloss formula)
- *Reaction Rate (average over length of pipe)
- *Water Quality (average over length of pipe)

The items marked with asterisks are computed quantities whose values will only be available if a successful analysis has been run on the network (see Chapter 8 – Analyzing a Network).

7.2 Setting the Map's Dimensions

The physical dimensions of the map must be defined so that map coordinates can be properly scaled to the computer's video display. To set the map's dimensions:

1. Select **View >> Dimensions**.

2. Enter new dimension information into the Map Dimensions dialog that appears (see Figure 7.1) or click the **Auto-Size** button to have EPANET compute dimensions based on the coordinates of objects currently included in the network.

3. Click the **OK** button to re-size the map.

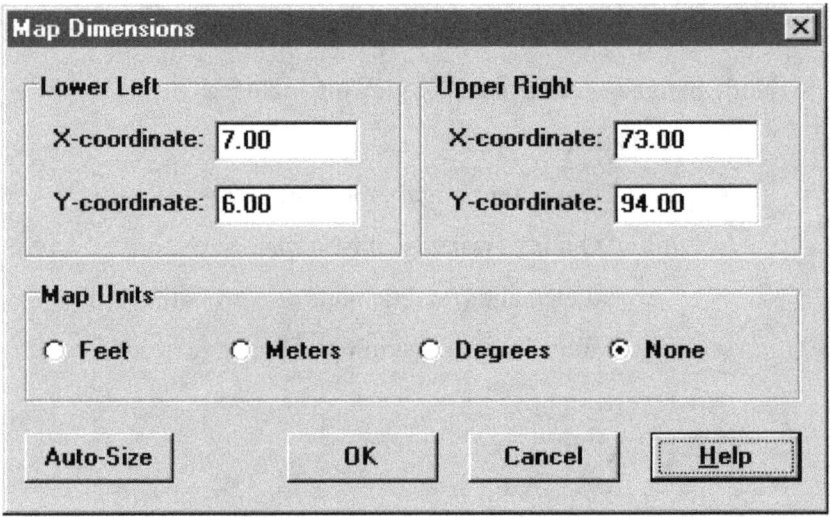

Figure 7.1 Map Dimensions Dialog

The information provided in the Map Dimensions dialog consists of the following:

Item	Description
Lower Left Coordinates	The X and Y coordinates of the lower left point on the map.
Upper Right Coordinates	The X and Y coordinates of the upper right point on the map.
Map Units	Units used to measure distances on the map. Choices are Feet, Meters, Degrees, and None (i.e., arbitrary units).

Note: If you are going to use a backdrop map with automatic pipe length calculation, then it is recommended that you set the map dimensions immediately after creating a new project. Map distance units can be different from pipe length units. The latter (feet or meters) depend on whether flow rates are expressed in US or metric units. EPANET will automatically convert units if necessary.

7.3 Utilizing a Backdrop Map

EPANET can display a backdrop map behind the pipe network map. The backdrop map might be a street map, utility map, topographic map, site development plan, or any other picture or drawing that might be useful. For example, using a street map would simplify the process of adding pipes to the network since one could essentially digitize the network's nodes and links directly on top of it.

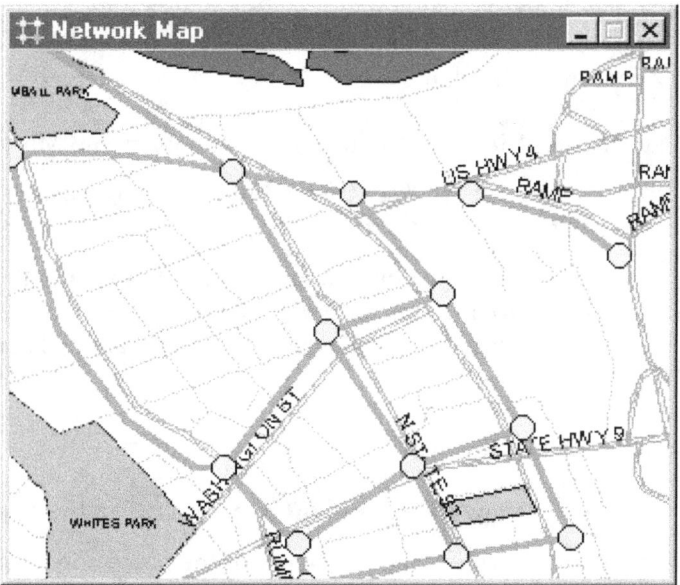

The backdrop map must be a Windows enhanced metafile or bitmap created outside of EPANET. Once imported, its features cannot be edited, although its scale and extent will change as the map window is zoomed and panned. For this reason metafiles work better than bitmaps since they will not loose resolution when re-scaled. Most CAD and GIS programs have the ability to save their drawings and maps as metafiles.

Selecting **View >> Backdrop** from the Menu Bar will display a sub-menu with the following commands:

- **Load** (loads a backdrop map file into the project)
- **Unload** (unloads the backdrop map from the project)
- **Align** (aligns the pipe network with the backdrop)
- **Show/Hide** (toggles the display of the backdrop on and off)

When first loaded, the backdrop image is placed with its upper left corner coinciding with that of the network's bounding rectangle. The backdrop can be re-positioned relative to the network map by selecting **View >> Backdrop >> Align**. This allows an outline of the pipe network to be moved across the backdrop (by moving the mouse with the left button held down) until one decides that it lines up properly with the backdrop. The name of the backdrop file and its current alignment are saved along with the rest of a project's data whenever the project is saved to file.

For best results in using a backdrop map:

- Use a metafile, not a bitmap.
- Dimension the network map so that its bounding rectangle has the same aspect ratio (width-to-height ratio) as the backdrop.

7.4 Zooming the Map

To Zoom In on the map:

1. Select **View >> Zoom In** or click [icon] on the Map Toolbar.
2. To zoom in 100%, move the mouse to the center of the zoom area and click the left button.
3. To perform a custom zoom, move the mouse to the upper left corner of the zoom area and with the left button pressed down, draw a rectangular outline around the zoom area. Then release the left button.

To Zoom Out on the map:

1. Select **View >> Zoom Out** or click [icon] on the Map Toolbar.
2. Move the mouse to the center of the new zoom area and click the left button.
3. The map will be returned to its previous zoom level.

7.5 Panning the Map

To pan the map across the Map window:

1. Select **View >> Pan** or click [icon] on the Map Toolbar.
2. With the left button held down over any point on the map, drag the mouse in the direction you wish to pan in.
3. Release the mouse button to complete the pan.

To pan using the Overview Map (which is described in Section 7.7 below):

1. If not already visible, bring up the Overview Map by selecting **View >> Overview Map.**
2. Position the mouse within the zoom window displayed on the Overview Map.
3. With the left button held down, drag the zoom window to a new position.
4. Release the mouse button and the main map will be panned to an area corresponding to that of the Overview Map's zoom window.

7.6 Finding an Object

To find a node or link on the map whose ID label is known:

1. Select **View >> Find** or click ![icon] on the Standard Toolbar.
2. In the Map Finder dialog box that appears, select **Node** or **Link** and enter an ID label.
3. Click **Find**.

If the node/link exists it will be highlighted on the map and in the Browser. If the map is currently zoomed in and the node/link falls outside the current map boundaries, the map will be panned so that the node/link comes into view. The Map Finder dialog will also list the ID labels of the links that connect to a found node or the nodes attached to a found link.

To find a listing of all nodes that serve as water quality sources:

1. Select **View >> Find** or click ![icon] on the Standard Toolbar.
2. In the Map Finder dialog box that appears, select **Sources**.
3. Click **Find**.

The ID labels of all water quality source nodes will be listed in the Map Finder. Clicking on any ID label will highlight that node on the map.

7.7 Map Legends

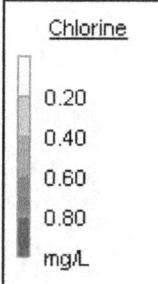

There are three types of map legends that can be displayed. The Node and Link Legends associate a color with a range of values for the current parameter being viewed on the map. The Time Legend displays the clock time of the simulation time period being viewed. To display or hide any of these legends check or uncheck the legend from the **View >> Legends** menu or right-click over the map and do the same from the popup menu that appears. Double-clicking the mouse over it can also hide a visible legend.

To move a legend to another location:

1. Press the left mouse button over the legend.
2. With the button held down, drag the legend to its new location and release the button.

To edit the Node Legend:

1. Either select **View >> Legends >> Modify >> Node** or right-click on the legend if it is visible.

2. Use the Legend Editor dialog form that appears (see Figure 7.2) to modify the legend's colors and intervals.

A similar method is used to edit the Link Legend.

The Legend Editor (Figure 7.2) is used to set numerical ranges to which different colors are assigned for viewing a particular parameter on the network map. It works as follows:

- Numerical values, in increasing order, are entered in the edit boxes to define the ranges. Not all four boxes need to have values.

- To change a color, click on its color band in the Editor and then select a new color from the Color Dialog box that will appear.

- Click the **Equal Intervals** button to assign ranges based on dividing the range of the parameter at the current time period into equal intervals.

- Click the **Equal Quantiles** button to assign ranges so that there are equal numbers of objects within each range, based on values that exist at the current time period.

- The **Color Ramp** button is used to select from a list of built-in color schemes.

- The **Reverse Colors** button reverses the ordering of the current set of colors (the color in the lowest range becomes that of the highest range and so on).

- Check **Framed** if you want a frame drawn around the legend.

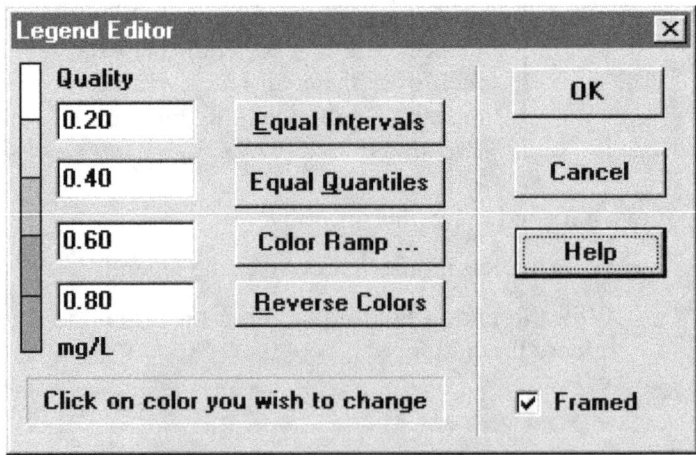

Figure 7.2 Legend Editor Dialog

7.8 Overview Map

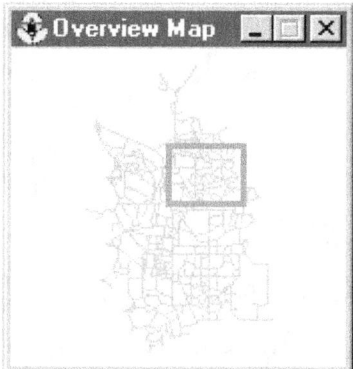

The Overview Map allows you to see where in terms of the overall system the main network map is currently focused. This zoom area is depicted by the rectangular boundary displayed on the Overview Map. As you drag this rectangle to another position the view within the main map will follow suit. The Overview Map can be toggled on and off by selecting **View >> Overview Map**. Clicking the mouse on its title bar will update its map image to match that of the main network map.

7.9 Map Display Options

There are several ways to bring up the Map Options dialog form (Figure 7.3) used to change the appearance of the Network Map:

- select **View >> Options**,
- click the Options button on the Standard Toolbar when the Map window has the focus,
- right-click on any empty portion of the map and select **Options** from the popup menu that appears.

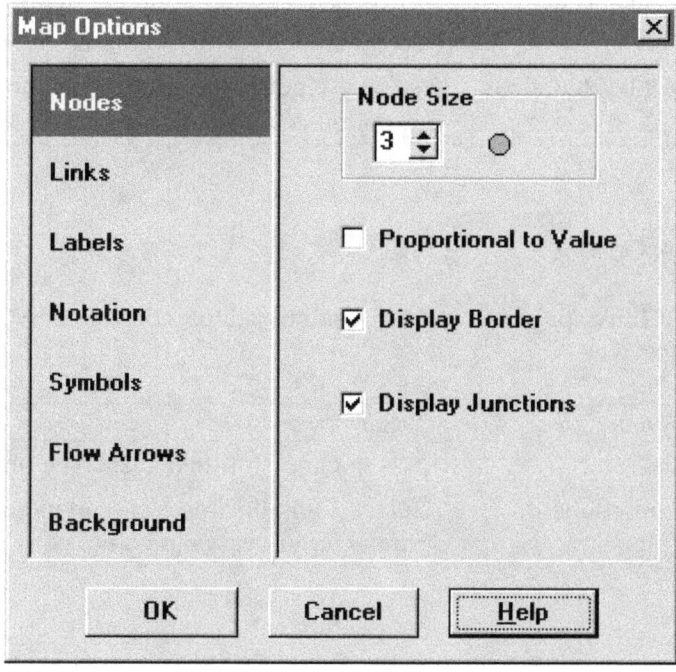

Figure 7.3 Map Options Dialog

The dialog contains a separate page, selected from the panel on the left side of the form, for each of the following display option categories:

- *Nodes* (controls size of nodes and making size be proportional to value)
- *Links* (controls thickness of links and making thickness be proportional to value)
- Labels (turns display of map labels on/off)
- *Notation* (displays or hides node/link ID labels and parameter values)
- *Symbols* (turns display of tank, pump, valve symbols on/off)
- *Flow Arrows* (selects visibility and style of flow direction arrows)
- *Background* (changes color of map's background)

Node Options

The Nodes page of the Map Options dialog controls how nodes are displayed on the Network Map.

Option	Description
Node Size	Selects node diameter
Proportional to Value	Select if node size should increase as the viewed parameter increases in value
Display Border	Select if a border should be drawn around each node (recommended for light-colored backgrounds)
Display Junctions	Displays junction nodes (all junctions will be hidden unless this option is checked).

Link Options

The Links page of the Map Options dialog controls how links are displayed on the map.

Option	Description
Link Size	Sets thickness of links displayed on map
Proportional to Value	Select if link thickness should increase as the viewed parameter increases in value

Label Options

The Labels page of the Map Options dialog controls how labels are displayed on the map.

Option	Description
Display Labels	Displays map labels (labels will be hidden unless this option is checked)
Use Transparent Text	Displays label with a transparent background (otherwise an opaque background is used)
At Zoom Of	Selects minimum zoom at which labels should be displayed; labels will be hidden at zooms smaller than this unless they are meter labels

Notation Options

The Notation page of the Map Options dialog form determines what kind of annotation is provided alongside of the nodes and links of the map.

Option	Description
Display Node IDs	Displays node ID labels
Display Node Values	Displays value of current node parameter being viewed
Display Link IDs	Displays link ID labels
Display Link Values	Displays values of current link parameter being viewed
Use Transparent Text	Displays text with a transparent background (otherwise an opaque background is used)
At Zoom Of	Selects minimum zoom at which notation should be displayed; all notation will be hidden at zooms smaller than this

Note: Values of the current viewing parameter at only specific nodes and links can be displayed by creating Map Labels with meters for those objects. See Sections 6.2 and 6.4 as well as Table 6.7.

Symbol Options

The Symbols page of the Map Options dialog determines which types of objects are represented with special symbols on the map.

Option	Description
Display Tanks	Displays tank symbols
Display Pumps	Displays pump symbols
Display Valves	Displays valve symbols
Display Emitters	Displays emitter symbols
Display Sources	Displays **+** symbol for water quality sources
At Zoom Of	Selects minimum zoom at which symbols should be displayed; symbols will be hidden at zooms smaller than this

Flow Arrow Options

The Flow Arrows page of the Map Options dialog controls how flow-direction arrows are displayed on the network map.

Option	Description
Arrow Style	Selects style (shape) of arrow to display (select None to hide arrows)
Arrow Size	Sets arrow size
At Zoom Of	Selects minimum zoom at which arrows should be displayed; arrows will be hidden at zooms smaller than this

Note: Flow direction arrows will only be displayed after a network has been successfully analyzed (see Section 8.2 Running an Analysis).

Background Options

The Background page of the Map Options dialog offers a selection of colors used to paint the map's background with.

CHAPTER 8 - ANALYZING A NETWORK

After a network has been suitably described, its hydraulic and water quality behavior can be analyzed. This chapter describes how to specify options to use in the analysis, how to run the analysis and how to troubleshoot problems that might have occurred with the analysis.

8.1 Setting Analysis Options

There are five categories of options that control how EPANET analyzes a network: Hydraulics, Quality, Reactions, Times, and Energy. To set any of these options:

1. Select the Options category from the Data Browser or select **Project >> Analysis Options** from the menu bar.
2. Select Hydraulics, Quality, Reactions, Times, or Energy from the Browser.
3. If the Property Editor is not already visible, click the Browser's Edit button (or hit the **Enter** key).
4. Edit your option choices in the Property Editor.

As you are editing a category of options in the Property Editor you can move to the next or previous category by simply hitting the **Page Down** or **Page Up** keys, respectively.

Hydraulic Options

Hydraulic options control how the hydraulic computations are carried out. They consist of the following items:

Option	Description
Flow Units	Units in which nodal demands and link flow rates are expressed. Choosing units in gallons, cubic feet, or acre-feet implies that the units for all other network quantities are Customary US. Selecting liters or cubic meters causes all other units to be SI metric. Use caution when changing flow units as it might affect all other data supplied to the project. (See Appendix A, Units of Measurement.)
Headloss Formula	Formula used to compute headloss as a function of flow rate in a pipe. Choices are: • Hazen-Williams • Darcy-Weisbach • Chezy-Manning Because each formula measures pipe roughness differently, switching formulas might require that all pipe roughness coefficients be updated.

Specific Gravity	Ratio of the density of the fluid being modeled to that of water at 4 deg. C (unitless).
Relative Viscosity	Ratio of the kinematic viscosity of the fluid to that of water at 20 deg. C (1.0 centistokes or 0.94 sq ft/day) (unitless).
Maximum Trials	Maximum number of trials used to solve the nonlinear equations that govern network hydraulics at a given point in time. Suggested value is 40.
Accuracy	Convergence criterion used to signal that a solution has been found to the nonlinear equations that govern network hydraulics. Trials end when the sum of all flow changes divided by the sum of all link flows is less than this number. Suggested value is 0.001.
If Unbalanced	Action to take if a hydraulic solution is not found within the maximum number of trials. Choices are STOP to stop the simulation at this point or CONTINUE to use another 10 trials, with no link status changes allowed, in an attempt to achieve convergence.
Default Pattern	ID label of a time pattern to be applied to demands at those junctions where no time pattern is specified. If no such pattern exists then demands will not vary at these locations.
Demand Multiplier	Global multiplier applied to all demands to make total system consumption vary up or down by a fixed amount. E.g., 2.0 doubles all demands, 0.5 halves them, and 1.0 leaves them as is.
Emitter Exponent	Power to which pressure is raised when computing the flow through an emitter device. The textbook value for nozzles and sprinklers is ½. This may not apply to pipe leakage. Consult the discussion of Emitters in Section 3.1 for more details.
Status Report	Amount of status information to report after an analysis is made. Choices are: • NONE (no status reporting) • YES (normal status reporting – lists all changes in link status throughout the simulation) • FULL (full reporting – normal reporting plus the convergence error from each trial of the hydraulic analysis made in each time period) Full status reporting is only useful for debugging purposes.

Note: Choices for Hydraulic Options can also be set from the **Project >> Defaults** menu and saved for use with all future projects (see Section 5.2).

Water Quality Options

Water Quality Options control how the water quality analysis is carried out. They consist of the following:

Option	Description
Parameter	Type of water quality parameter being modeled. Choices include: • NONE (no quality analysis), • CHEMICAL (compute chemical concentration), • AGE (compute water age), • TRACE (trace the percent of flow originating from a specific node). In lieu of CHEMICAL, you can enter the actual name of the chemical being modeled (e.g., Chlorine).
Mass Units	Mass units used to express concentration. Choices are mg/L or µg/L. Units for Age and Trace analyses are fixed at hours and percent, respectively.
Relative Diffusivity	Ratio of the molecular diffusivity of the chemical being modeled to that of chlorine at 20 deg. C (0.00112 sq ft/day). Use 2 if the chemical diffuses twice as fast as chlorine, 0.5 if half as fast, etc. Applies only when modeling mass transfer for pipe wall reactions. Set to zero to ignore mass transfer effects.
Trace Node	ID label of the node whose flow is being traced. Applies only to flow tracing analyses.
Quality Tolerance	Smallest change in quality that will cause a new parcel of water to be created in a pipe. A typical setting might be 0.01 for chemicals measured in mg/L as well as water age and source tracing.

Note: The Quality Tolerance determines when the quality of one parcel of water is essentially the same as another parcel. For chemical analysis this might be the detection limit of the procedure used to measure the chemical, adjusted by a suitable factor of safety. Using too large a value for this tolerance might affect simulation accuracy. Using too small a value will affect computational efficiency. Some experimentation with this setting might be called for.

Reaction Options

Reaction Options set the types of reactions that apply to a water quality analysis. They include the following:

Option	Description
Bulk Reaction Order	Power to which concentration is raised when computing a bulk flow reaction rate. Use 1 for first-order reactions, 2 for second-order reactions, etc. Use any negative number for Michaelis-Menton kinetics. If no global or pipe-specific bulk reaction coefficients are assigned then this option is ignored.
Wall Reaction Order	Power to which concentration is raised when computing a bulk flow reaction rate. Choices are FIRST (1) for first-order reactions or ZERO (0) for constant rate reactions. If no global or pipe-specific wall reaction coefficients are assigned then this option is ignored.
Global Bulk Coefficient	Default bulk reaction rate coefficient (K_b) assigned to all pipes. This global coefficient can be overridden by editing this property for specific pipes. Use a positive number for growth, a negative number for decay, or 0 if no bulk reaction occurs. Units are concentration raised to the (1-n) power divided by days, where n is the bulk reaction order.
Global Wall Coefficient	Wall reaction rate coefficient (K_w) assigned to all pipes. Can be overridden by editing this property for specific pipes. Use a positive number for growth, a negative number for decay, or 0 if no wall reaction occurs. Units are ft/day (US) or m/day (SI) for first-order reactions and mass/sq ft/day (US) or mass/sq m/day (SI) for zero-order reactions.
Limiting Concentration	Maximum concentration that a substance can grow to or minimum value it can decay to. Bulk reaction rates will be proportional to the difference between the current concentration and this value. See discussion of Bulk Reactions in Section 3.4 for more details. Set to zero if not applicable.
Wall Coefficient Correlation	Factor correlating wall reaction coefficient to pipe roughness. See discussion of Wall Reactions in Section 3.4 for more details. Set to zero if not applicable.

Times Options

Times options set values for the various time steps used in an extended period simulation. These are listed below (times can be entered as decimal hours or in hours:minutes notation):

Option	Description
Total Duration	Total length of a simulation in hours. Use 0 to run a single period (snapshot) hydraulic analysis.
Hydraulic Time Step	Time interval between re-computation of system hydraulics. Normal default is 1 hour.
Quality Time Step	Time interval between routing of water quality constituent. Normal default is 5 minutes (0:05 hours).
Pattern Time Step	Time interval used with all time patterns. Normal default is 1 hour.
Pattern Start Time	Hours into all time patterns at which the simulation begins (e.g., a value of 2 means that the simulation begins with all time patterns starting at their second hour). Normal default is 0.
Reporting Time Step	Time interval between times at which computed results are reported. Normal default is 1 hour.
Report Start Time	Hours into simulation at which computed results begin to be reported. Normal default is 0.
Starting Time of Day	Clock time (e.g., 7:30 am, 10:00 pm) at which simulation begins. Default is 12:00 am (midnight).
Statistic	Type of statistical processing used to summarize the results of an extended period simulation. Choices are: • NONE (results reported at each reporting time step) • AVERAGE (time-averaged results reported) • MINIMUM (minimum value results reported) • MAXIMUM (maximum value results reported) • RANGE (difference between maximum and minimum results reported) Statistical processing is applied to all node and link results obtained between the Report Start Time and the Total Duration.

Note: To run a single-period hydraulic analyses (also called a snapshot analysis) enter 0 for Total Duration. In this case entries for all of the other time options, with the exception of Starting Time of Day, are not used. Water quality analyses always require that a non-zero Total Duration be specified.

Energy Options

Energy Analysis Options provide default values used to compute pumping energy and cost when no specific energy parameters are assigned to a given pump. They consist of the following:

Option	Description
Pump Efficiency (%)	Default pump efficiency.
Energy Price per Kwh	Price of energy per kilowatt-hour. Monetary units are not explicitly represented.
Price Pattern	ID label of a time pattern used to represent variations in energy price with time. Leave blank if not applicable.
Demand Charge	Additional energy charge per maximum kilowatt usage.

8.2 Running an Analysis

To run a hydraulic/water quality analysis:

1. Select **Project >> Run Analysis** or click ⚡ on the Standard Toolbar.
2. The progress of the analysis will be displayed in a Run Status window.
3. Click **OK** when the analysis ends.

If the analysis runs successfully the 🚰 icon will appear in the Run Status section of the Status Bar at the bottom of the EPANET workspace. Any error or warning messages will appear in a Status Report window. If you edit the properties of the network after a successful run has been made, the faucet icon changes to a broken faucet indicating that the current computed results no longer apply to the modified network.

8.3 Troubleshooting Results

EPANET will issue specific Error and Warning messages when problems are encountered in running a hydraulic/water quality analysis (see Appendix B for a complete listing). The most common problems are discussed below.

Pumps Cannot Deliver Flow or Head

EPANET will issue a warning message when a pump is asked to operate outside the range of its pump curve. If the pump is required to deliver more head than its shutoff head, EPANET will close the pump down. This might lead to portions of the network becoming disconnected from any source of water.

Network is Disconnected

EPANET classifies a network as being disconnected if there is no way to provide water to all nodes that have demands. This can occur if there is no path of open links between a junction with demand and either a reservoir, a tank, or a junction with a negative demand. If the problem is caused by a closed link EPANET will still compute a hydraulic solution (probably with extremely large negative pressures) and attempt to identify the problem link in its Status Report. If no connecting link(s) exist EPANET will be unable to solve the hydraulic equations for flows and pressures and will return an Error 110 message when an analysis is made. Under an extended period simulation it is possible for nodes to become disconnected as links change status over time.

Negative Pressures Exist

EPANET will issue a warning message when it encounters negative pressures at junctions that have positive demands. This usually indicates that there is some problem with the way the network has been designed or operated. Negative pressures can occur when portions of the network can only receive water through links that have been closed off. In such cases an additional warning message about the network being disconnected is also issued.

System Unbalanced

A System Unbalanced condition can occur when EPANET cannot converge to a hydraulic solution in some time period within its allowed maximum number of trials. This situation can occur when valves, pumps, or pipelines keep switching their status from one trial to the next as the search for a hydraulic solution proceeds. For example, the pressure limits that control the status of a pump may be set too close together. Or a pump's head curve might be too flat causing it to keep shutting on and off.

To eliminate the unbalanced condition one can try to increase the allowed maximum number of trials or loosen the convergence accuracy requirement. Both of these parameters are set with the project's Hydraulic Options. If the unbalanced condition persists, then another hydraulic option, labeled "If Unbalanced", offers two ways to handle it. One is to terminate the entire analysis once the condition is encountered. The other is to continue seeking a hydraulic solution for another 10 trials with the status of all links frozen to their current values. If convergence is achieved then a warning message is issued about the system possibly being unstable. If convergence is not achieved then a "System Unbalanced" warning message is issued. In either case, the analysis will proceed to the next time period.

If an analysis in a given time period ends with the system unbalanced then the user should recognize that the hydraulic results produced for this time period are inaccurate. Depending on circumstances, such as errors in flows into or out of storage tanks, this might affect the accuracy of results in all future periods as well.

Hydraulic Equations Unsolvable

Error 110 is issued if at some point in an analysis the set of equations that model flow and energy balance in the network cannot be solved. This can occur when some portion of a system demands water but has no links physically connecting it to any source of water. In such a case EPANET will also issue warning messages about nodes being disconnected. The equations might also be unsolvable if unrealistic numbers were used for certain network properties.

CHAPTER 9 - VIEWING RESULTS

This chapter describes the different ways in which the results of an analysis as well as the basic network input data can be viewed. These include different map views, graphs, tables, and special reports.

9.1 Viewing Results on the Map

There are several ways in which database values and results of a simulation can be viewed directly on the Network Map:

- For the current settings on the Map Browser (see Section 4.6), the nodes and links of the map will be colored according to the color-coding used in the Map Legends (see Section 7.6). The map's coloring will be updated as a new time period is selected in the Browser.

- When the Flyover Map Labeling program preference is selected (see Section 4.9), moving the mouse over any node or link will display its ID label and the value of the current viewing parameter for that node or link in a hint-style box.

- ID labels and viewing parameter values can be displayed next to all nodes and/or links by selecting the appropriate options on the Notation page of the Map Options dialog form (see Section 7.8).

- Nodes or links meeting a specific criterion can be identified by submitting a Map Query (see below).

- You can animate the display of results on the network map either forward or backward in time by using the Animation buttons on the Map Browser. Animation is only available when a node or link viewing parameter is a computed value (e.g., link flow rate can be animated but diameter cannot).

- The map can be printed, copied to the Windows clipboard, or saved as a DXF file or Windows metafile.

Submitting a Map Query

A Map Query identifies nodes or links on the network map that meet a specific criterion (e.g., nodes with pressure less than 20 psi, links with velocity above 2 ft/sec, etc.). See Figure 9.1 for an example. To submit a map query:

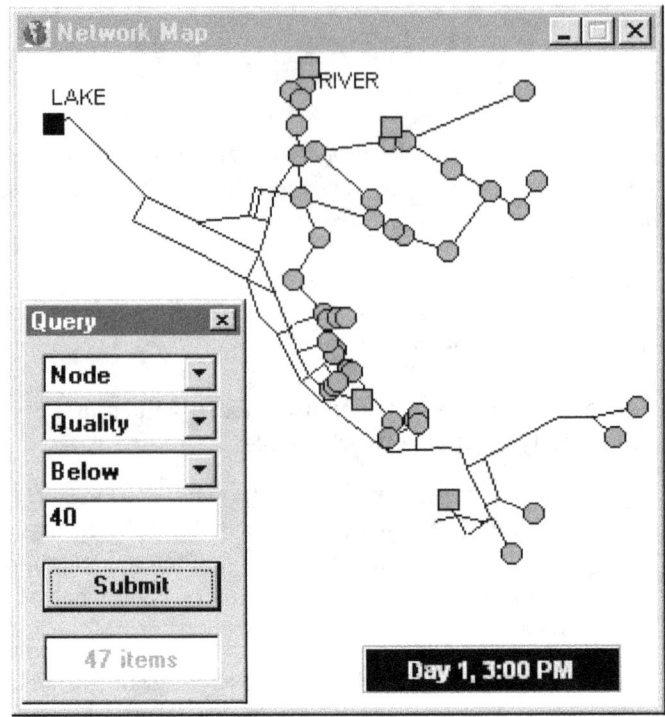

Figure 9.1 Results of a Map Query

1. Select a time period in which to query the map from the Map Browser.

2. Select **View >> Query** or click [icon] on the Map Toolbar.

3. Fill in the following information in the Query dialog form that appears:
    - Select whether to search for Nodes or Links
    - Select a parameter to compare against
    - Select **Above**, **Below**, or **Equals**
    - Enter a value to compare against

4. Click the **Submit** button. The objects that meet the criterion will be highlighted on the map.

5. As a new time period is selected in the Browser, the query results are automatically updated.

6. You can submit another query using the dialog box or close it by clicking the button in the upper right corner.

After the Query box is closed the map will revert back to its original display.

9.2 Viewing Results with a Graph

Analysis results, as well as some design parameters, can be viewed using several different types of graphs. Graphs can be printed, copied to the Windows clipboard, or saved as a data file or Windows metafile. The following types of graphs can be used to view values for a selected parameter (see Figure 9.2 for examples of each):

Type of Plot	Description	Applies To
Time Series Plot	Plots value versus time	Specific nodes or links over all time periods
Profile Plot	Plots value versus distance	A list of nodes at a specific time
Contour Plot	Shows regions of the map where values fall within specific intervals	All nodes at a specific time
Frequency Plot	Plots value versus fraction of objects at or below the value	All nodes or links at a specific time
System Flow	Plots total system production and consumption versus time	Water demand for all nodes over all time periods

Note: When only a single node or link is graphed in a Time Series Plot the graph will also display any measured data residing in a Calibration File that has been registered with the project (see Section 5.3).

To create a graph:

1. Select **Report >> Graph** or click ▦ on the Standard Toolbar.
2. Fill in the choices on the Graph Selection dialog box that appears.
3. Click **OK** to create the graph.

The Graph Selection dialog, as pictured in Figure 9.3, is used to select a type of graph and its contents to display. The choices available in the dialog consist of the following:

Item	Description
Graph Type	Selects a graph type
Parameter	Selects a parameter to graph
Time Period	Selects a time period to graph (does not apply to Time Series plots)
Object Type	Selects either Nodes or Links (only Nodes can be graphed on Profile and Contour plots)
Items to Graph	Selects items to graph (applies only to Time Series and Profile plots)

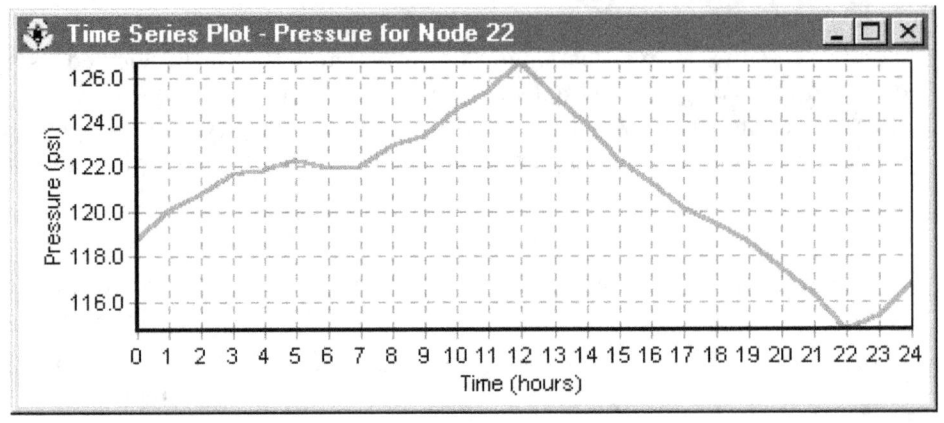

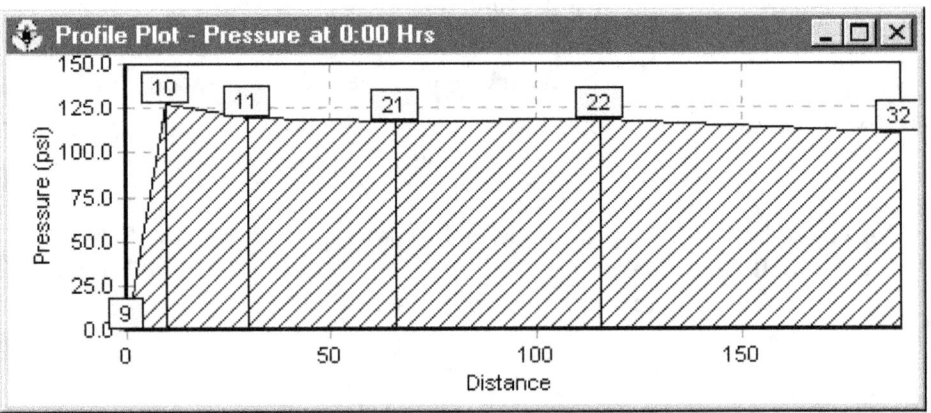

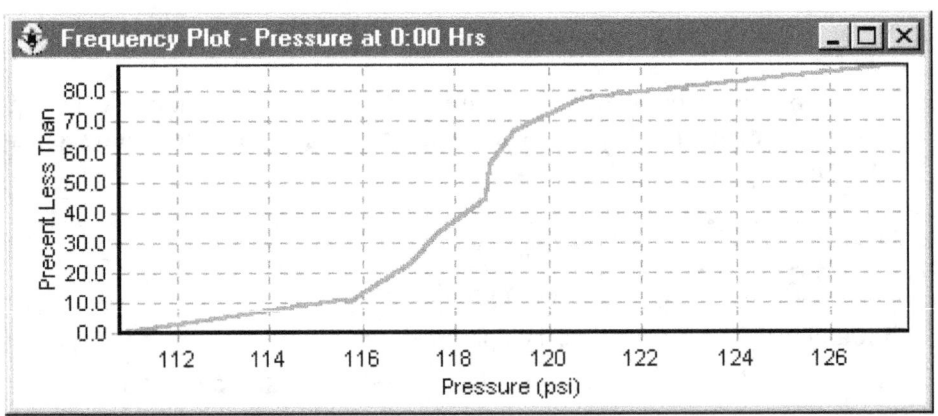

Figure 9.2 Examples of Different Types of Graphs

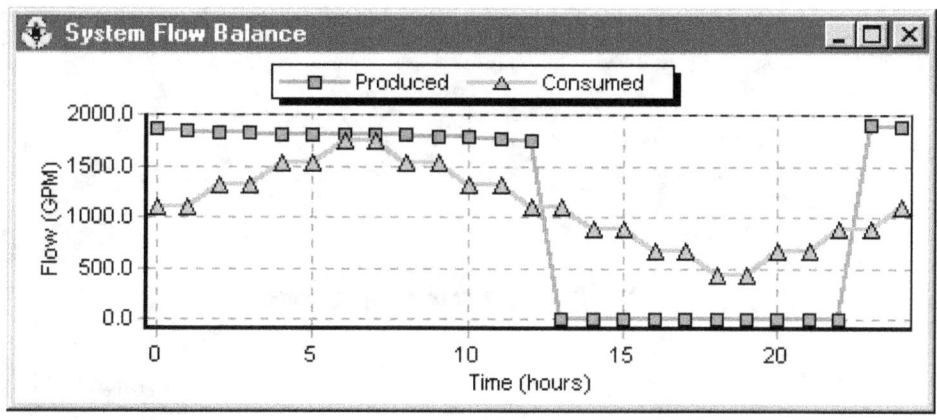

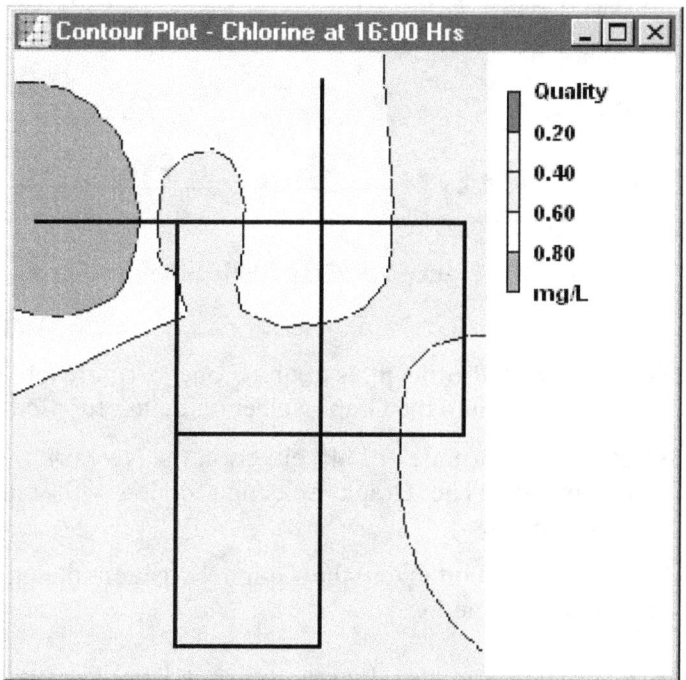

Figure 9.2 Continued From Previous Page

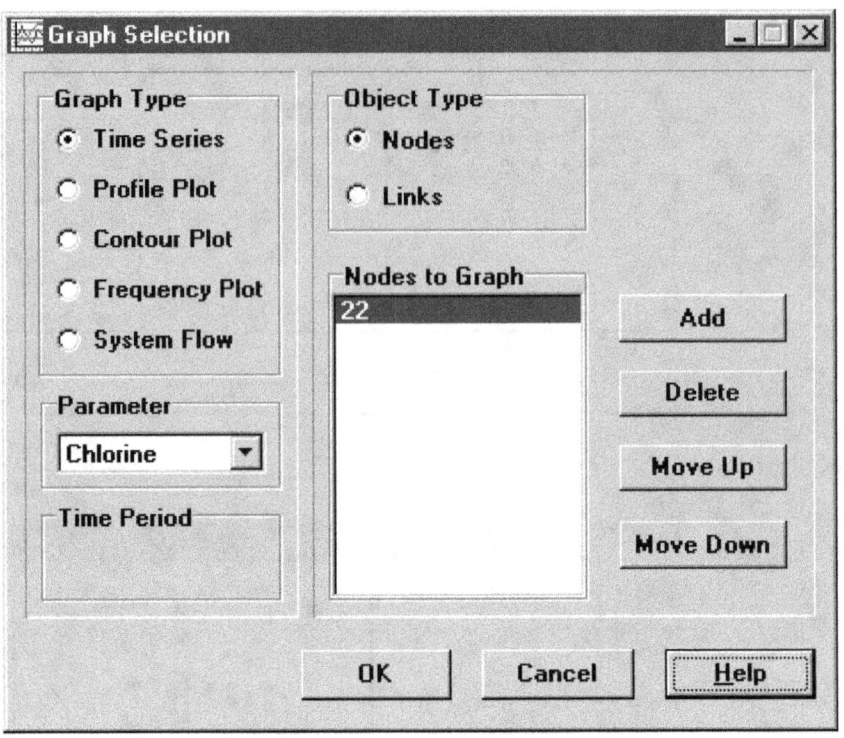

Figure 9.3 Graph Selection Dialog

Time Series plots and Profile plots require one or more objects be selected for plotting. To select items into the Graph Selection dialog for plotting:

1. Select the object (node or link) either on the Network Map or on the Data Browser. (The Graph Selection dialog will remain visible during this process).

2. Click the **Add** button on the Graph Selection dialog to add the selected item to the list.

In place of Step 2, you can also drag the object's label from the Data Browser onto the Form's title bar or onto the Items to Graph list box.

The other buttons on the **Graph Selection** dialog form are used as follows:

Button	Purpose
Load (Profile Plot Only)	Loads a previously saved list of nodes
Save (Profile Plot Only)	Saves current list of nodes to file
Delete	Deletes selected item from list
Move Up	Moves selected item on list up one position
Move Down	Moves selected item on list down one position

To customize the appearance of a graph:

1. Make the graph the active window (click on its title bar).
2. Select **Report** >> **Options**, or click ⬜ on the Standard Toolbar, or right-click on the graph.
3. For a Time Series, Profile, Frequency or System Flow plot, use the resulting Graph Options dialog (Figure 9.4) to customize the graph's appearance.
4. For a Contour plot use the resulting Contour Options dialog to customize the plot.

Note: A Time Series, Profile, or Frequency plot can be zoomed by holding down the **Ctrl** key while drawing a zoom rectangle with the mouse's left button held down. Drawing the rectangle from left to right zooms in, drawing from right to left zooms out. The plot can also be panned in any direction by holding down the **Ctrl** key and moving the mouse across the plot with the right button held down.

The Graph Options dialog form (Figure 9.4) is used to customize the appearance of an X-Y graph. To use the dialog box:

1. Select from among the five tabbed pages that cover the following categories of options:

   - General
   - Horizontal Axis
   - Vertical Axis
   - Legend
   - Series

2. Check the **Default** box if you wish to use the current settings as defaults for all new graphs as well.
3. Select **OK** to accept your selections.

The items contained on each page of the Graph Options dialog are as follows:

General Page

Option	*Description*
Panel Color	Color of the panel which surrounds the graph's plotting area
Background Color	Color of graph's plotting area
View in 3D	Check if graph should be drawn in 3D
3D Effect Percent	Degree to which 3D effect is drawn
Main Title	Text of graph's main title
Font	Changes the font used for the main title

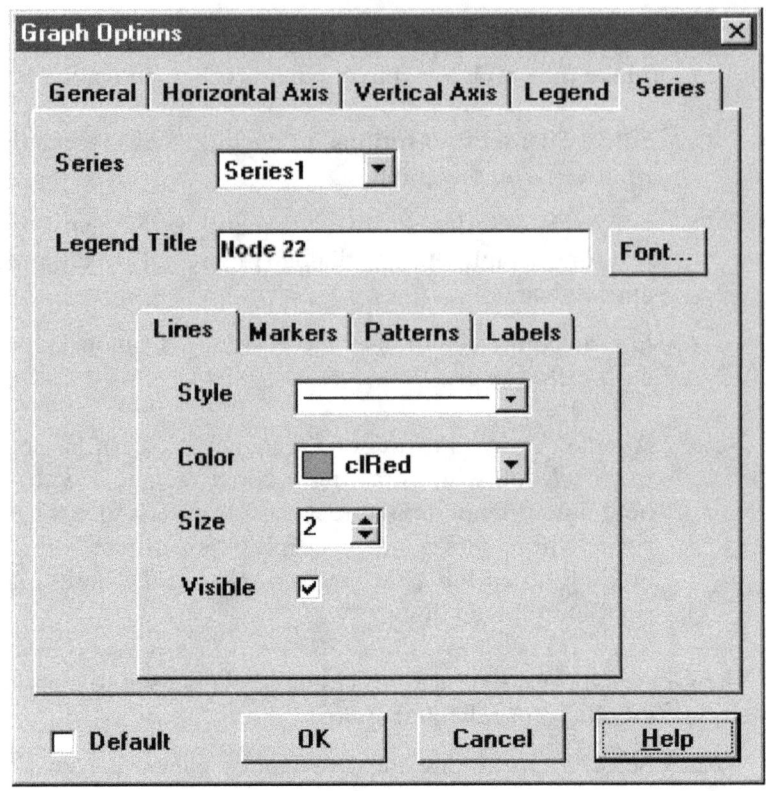

Figure 9.4 Graph Options Dialog

Horizontal and Vertical Axis Pages

Option	Description
Minimum	Sets minimum axis value (minimum data value is shown in parentheses). Can be left blank.
Maximum	Sets maximum axis value (maximum data value is shown in parentheses). Can be left blank.
Increment	Sets increment between axis labels. Can be left blank.
Auto Scale	If checked then Minimum, Maximum, and Increment settings are ignored.
Gridlines	Selects type of gridline to draw.
Axis Title	Text of axis title
Font	Click to select a font for the axis title.

Legend Page

Option	Description
Position	Selects where to place the legend.
Color	Selects color to use for legend background.
Symbol Width	Selects width to use (in pixels) to draw symbol portion of the legend.
Framed	Places a frame around the legend.
Visible	Makes the legend visible.

Series Page

The Series page (see Figure 9.4) of the Graph Options dialog controls how individual data series (or curves) are displayed on a graph. To use this page:

- Select a data series to work with from the Series combo box.
- Edit the title used to identify this series in the legend.
- Click the Font button to change the font used for the legend. (Other legend properties are selected on the Legend page of the dialog.)
- Select a property of the data series you would like to modify. The choices are:
    - Lines
    - Markers
    - Patterns
    - Labels

(Not all properties are available for some types of graphs.)

The data series properties that can be modified include the following:

Category	Option	Description
Lines	Style	Selects line style.
	Color	Selects line color.
	Size	Selects line thickness (only for solid line style).
	Visible	Determines if line is visible.
Markers	Style	Selects marker style.
	Color	Selects marker color.
	Size	Selects marker size.
	Visible	Determines if marker is visible.
Patterns	Style	Selects pattern style.
	Color	Selects pattern color.
	Stacking	Not used with EPANET.
Labels	Style	Selects what type of information is displayed in the label.
	Color	Selects the color of the label's background.
	Transparent	Determines if graph shows through label or not.
	Show Arrows	Determines if arrows are displayed on pie charts.
	Visible	Determines if labels are visible or not.

The Contour Options dialog form (Figure 9.5) is used to customize the appearance of a contour graph. A description of each option is provided below:

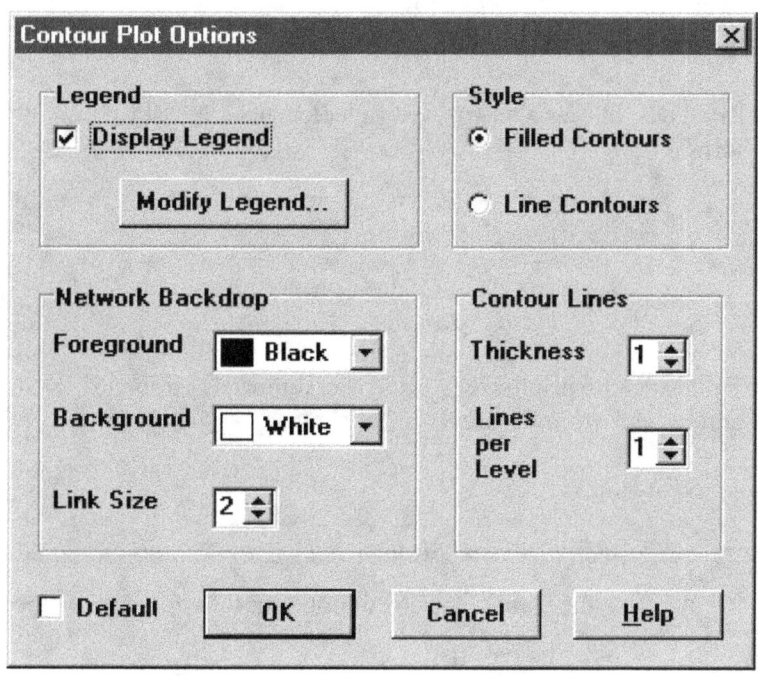

Figure 9.5 Contour Plot Options Dialog

Category	Option	Description
Legend	Display Legend	Toggles display of legend on/off
	Modify Legend	Changes colors and contour intervals
Network Backdrop	Foreground	Color of network image displayed on plot
	Background	Background color used for line contour plot
	Link Size	Thickness of lines used to display network
Style	Filled Contours	Plot uses colored area-filled contours
	Line Contours	Plot uses colored line contours
Contour Lines	Thickness	Thickness of lines used for contour intervals
	Lines per Level	Number of sub-contours per major contour level
Default		Saves choices as defaults for next contour plot

9.3 Viewing Results with a Table

EPANET allows you to view selected project data and analysis results in a tabular format:

- A <u>Network Table</u> lists properties and results for all nodes or links at a specific period of time.

- A <u>Time Series Table</u> lists properties and results for a specific node or link in all time periods.

Tables can be printed, copied to the Windows clipboard, or saved to file. An example table is shown in Figure 9.6.

To create a table:

1. Select **View >> Table** or click 🔲 on the Standard Toolbar.
2. Use the Table Options dialog box that appears to select:
   - the type of table
   - the quantities to display in each column
   - any filters to apply to the data

Node ID	Demand GPM	Head ft	Pressure psi	Chlorine mg/L
Junc 10	0.00	1010.67	130.28	1.00
Junc 11	210.00	992.42	122.37	0.85
Junc 12	210.00	980.17	121.40	0.78
Junc 13	140.00	977.08	122.23	0.30
Junc 21	210.00	977.24	120.13	0.74
Junc 22	280.00	976.29	121.88	0.49
Junc 23	210.00	975.76	123.82	0.30
Junc 31	140.00	970.32	117.13	0.53

Figure 9.6 Example Network Nodes Table

The Table Options dialog form has three tabbed pages as shown in Figure 9.7. All three pages are available when a table is first created. After the table is created, only the Columns and Filters tabs will appear. The options available on each page are as follows:

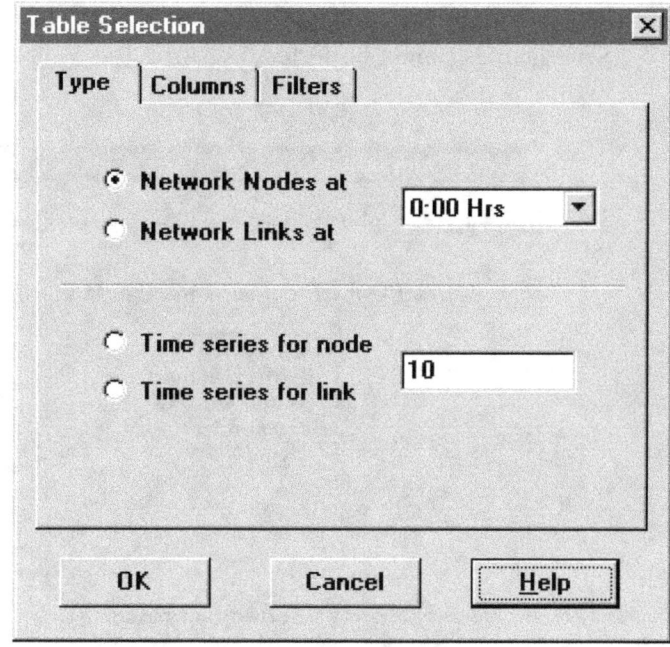

Figure 9.7 Table Selection Dialog

Type Page

The Type page of the Table Options dialog is used to select the type of table to create. The choices are:

- All network nodes at a specific time period
- All network links at a specific time period
- All time periods for a specific node
- All time periods for a specific link

Data fields are available for selecting the time period or node/link to which the table applies.

Columns Page

The Columns page of the Table Options dialog form (Figure 9.8) selects the parameters that are displayed in the table's columns.

- Click the checkbox next to the name of each parameter you wish to include in the table, or if the item is already selected, click in the box to deselect it. (The keyboard's Up and Down Arrow keys can be used to move between the parameter names, and the spacebar can be used to select/deselect choices).

- To sort a Network-type table with respect to the values of a particular parameter, select the parameter from the list and check off the **Sorted By** box at the bottom of the form. (The sorted parameter does not have to be selected as one of the columns in the table.) Time Series tables cannot be sorted.

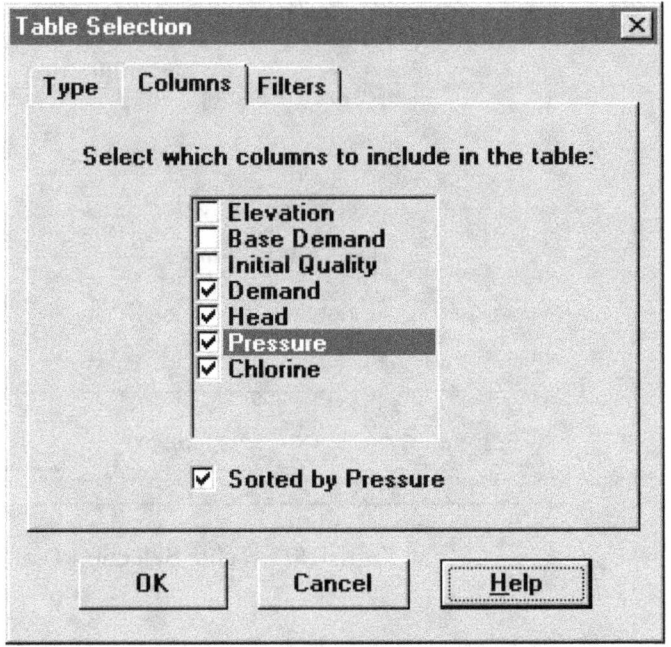

Figure 9.8 Columns Page of the Table Selection Dialog

Filters Page

The Filters page of the Table Options dialog form (Figure 9.9) is used to define conditions for selecting items to appear in a table. To filter the contents of a table:

- Use the controls at the top of the page to create a condition (e.g., Pressure Below 20).

- Click the **Add** button to add the condition to the list.

- Use the **Delete** button to remove a selected condition from the list.

Multiple conditions used to filter the table are connected by AND's. If a table has been filtered, a re-sizeable panel will appear at the bottom indicating how many items have satisfied the filter conditions.

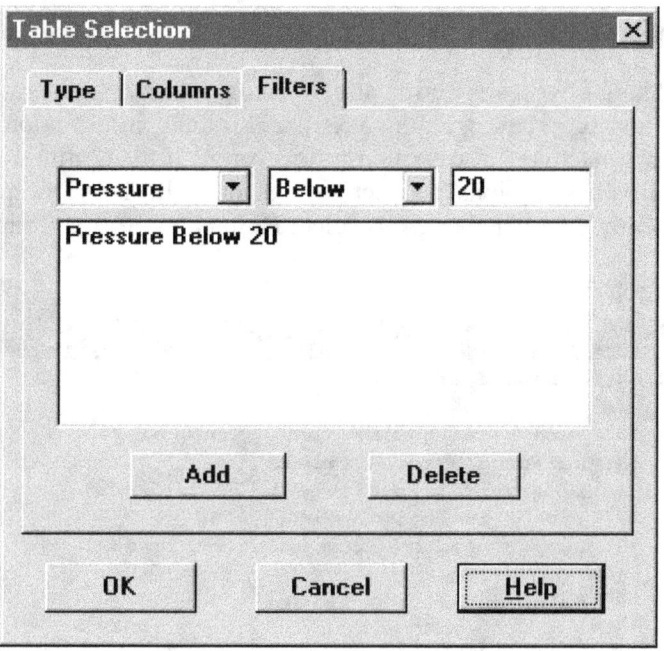

Figure 9.9 Filters Page of the Table Selection Dialog

Once a table has been created you can add/delete columns or sort or filter its data:

- Select **Report >> Options** or click [icon] on the Standard Toolbar or right-click on the table.
- Use the Columns and Filters pages of the Table Selection dialog form to modify your table.

9.4 Viewing Special Reports

In addition to graphs and tables, EPANET can generate several other specialized reports. These include:

- Status Report
- Energy Report
- Calibration Report
- Reaction Report
- Full Report

All of these reports can be printed, copied to a file, or copied to the Windows clipboard (the Full Report can only be saved to file.)

Status Report

EPANET writes all error and warning messages generated during an analysis to a Status Report (see Figure 9.10). Additional information on when network objects change status is also written to this report if the Status Report option in the project's Hydraulics Options was set to Yes or Full. To view a status report on the most recently completed analysis select **Report >> Status** from the main menu.

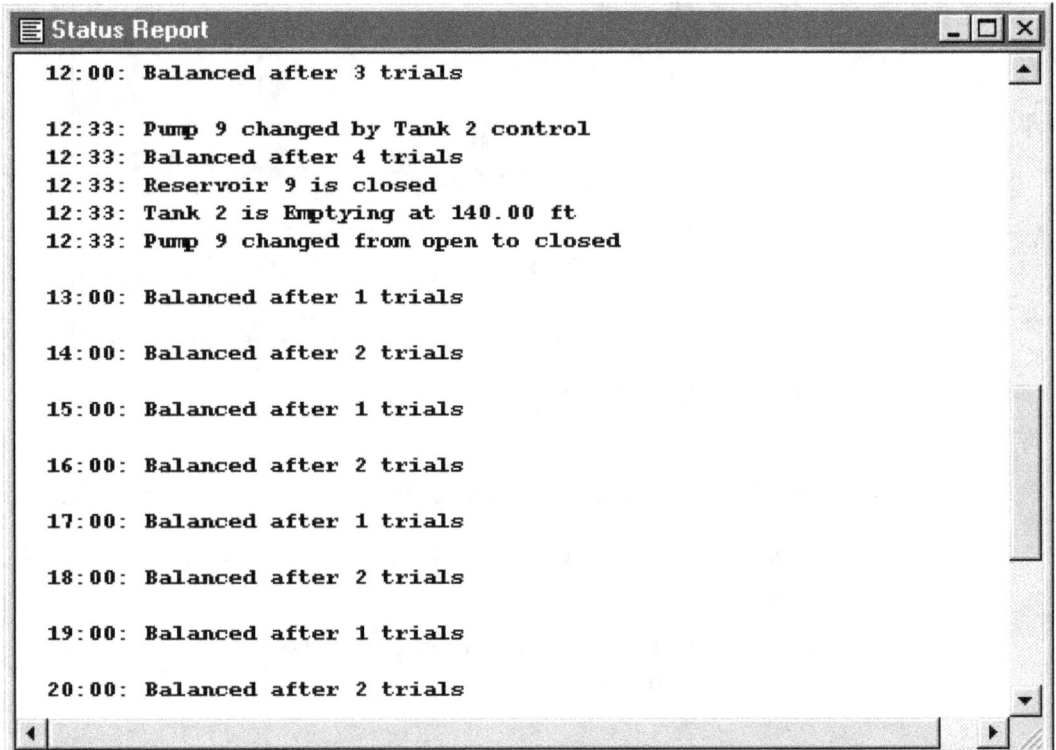

Figure 9.10 Excerpt from a Status Report

Energy Report

EPANET can generate an Energy Report that displays statistics about the energy consumed by each pump and the cost of this energy usage over the duration of a simulation (see Figure 9.11). To generate an Energy Report select **Report >> Energy** from the main menu. The report has two tabbed pages. One displays energy usage by pump in a tabular format. The second compares a selected energy statistic between pumps using a bar chart.

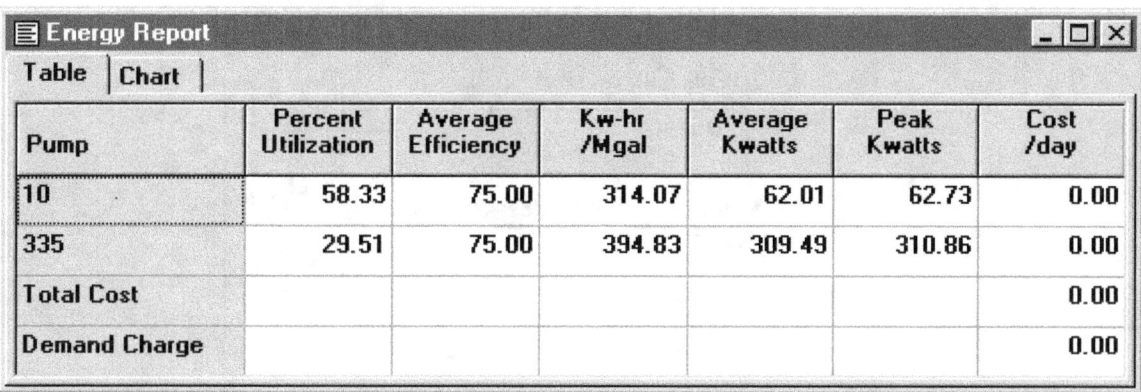

Figure 9.11 Example Energy Report

Calibration Report

A Calibration Report can show how well EPANET's simulated results match measurements taken from the system being modeled. To create a Calibration Report:

1. First make sure that Calibration Data for the quantity being calibrated has been registered with the project (see Section 5.3).
2. Select **Report >> Calibration** from the main menu.
3. In the Calibration Report Options form that appears (see Figure 9.12):
   - select a parameter to calibrate against
   - select the measurement locations to use in the report
4. Click **OK** to create the report.

After the report is created the Calibration Report Options form can be recalled to change report options by selecting **Report >> Options** or by clicking on the Standard Toolbar when the report is the current active window in EPANET's workspace.

A sample Calibration Report is shown in Figure 9.13. It contains three tabbed pages: Statistics, Correlation Plot, and Mean Comparisons.

Statistics Page

The Statistics page of a Calibration Report lists various error statistics between simulated and observed values at each measurement location and for the network as a whole. If a measured value at a location was taken at a time in-between the simulation's reporting time intervals then a simulated value for that time is found by interpolating between the simulated values at either end of the interval.

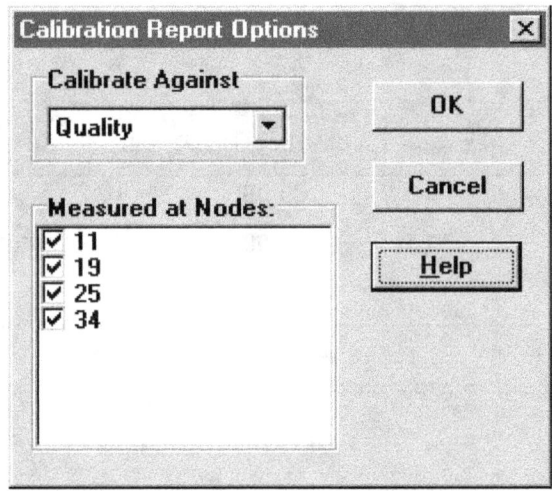

Figure 9.12 Calibration Report Options Dialog

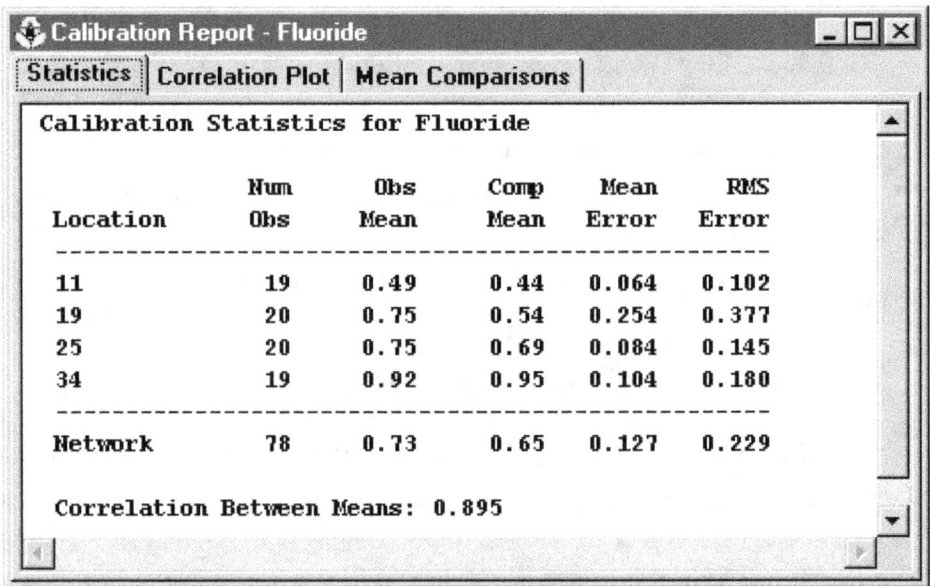

Figure 9.13 Example Calibration Report

The statistics listed for each measurement location are:

- Number of observations
- Mean of the observed values
- Mean of the simulated values
- Mean absolute error between each observed and simulated value
- Root mean square error (square root of the mean of the squared errors between the observed and simulated values).

These statistics are also provided for the network as a whole (i.e., all measurements and model errors pooled together). Also listed is the correlation between means (correlation coefficient between the mean observed value and mean simulated value at each location).

Correlation Plot Page

The Correlation Plot page of a Calibration Report displays a scatter plot of the observed and simulated values for each measurement made at each location. Each location is assigned a different color in the plot. The closer that the points come to the 45-degree angle line on the plot the closer is the match between observed and simulated values.

Mean Comparisons Page

The Mean Comparisons page of a Calibration Report presents a bar chart that compares the mean observed and mean simulated value for a calibration parameter at each location where measurements were taken.

Reaction Report

A Reaction Report, available when modeling the fate of a reactive water quality constituent, graphically depicts the overall average reaction rates occurring throughout the network in the following locations:

- the bulk flow
- the pipe wall
- within storage tanks.

A pie chart shows what percent of the overall reaction rate is occurring in each location. The chart legend displays the average rates in mass units per hour. A footnote on the chart shows the inflow rate of the reactant into the system.

The information in the Reaction Report can show at a glance what mechanism is responsible for the majority of growth or decay of a substance in the network. For example, if one observes that most of the chlorine decay in a system is occurring in the storage tanks and not at the walls of the pipes then one might infer that a corrective strategy of pipe cleaning and replacement will have little effect in improving chlorine residuals.

A Graph Options dialog box can be called up to modify the appearance of the pie chart by selecting **Report** >> **Options** or by clicking on the Standard Toolbar, or by right-clicking anywhere on the chart.

Full Report

When the ![icon] icon appears in the Run Status section of the Status Bar, a report of computed results for all nodes, links and time periods can be saved to file by selecting **Full** from the **Report** menu. This report, which can be viewed or printed outside of EPANET using any text editor or word processor, contains the following information:

- project title and notes
- a table listing the end nodes, length, and diameter of each link
- a table listing energy usage statistics for each pump
- a pair of tables for each time period listing computed values for each node (demand, head, pressure, and quality) and for each link (flow, velocity, headloss, and status).

This feature is useful mainly for documenting the final results of a network analysis on small to moderately sized networks (full report files for large networks analyzed over many time periods can easily consume dozens of megabytes of disk space). The other reporting tools described in this chapter are available for viewing computed results on a more selective basis.

CHAPTER 10 - PRINTING AND COPYING

This chapter describes how to print, copy to the Windows clipboard, or copy to file the contents of the currently active window in the EPANET workspace. This can include the network map, a graph, a table, a report, or the properties of an object selected from the Browser.

10.1 Selecting a Printer

To select a printer from among your installed Windows printers and set its properties:

1. Select **File >> Page Setup** from the main menu.
2. Click the **Printer** button on the Page Setup dialog that appears (see Figure 10.1).
3. Select a printer from the choices available in the combo box in the next dialog that appears.
4. Click the **Properties** button to select the printer's properties (which vary with choice of printer).
5. Click **OK** on each dialog box to accept your selections.

10.2 Setting the Page Format

To format the printed page:

1. Select **File >> Page Setup** from the main menu.
2. Use the Margins page of the Page Setup dialog form that appears (Figure 10.1) to:
   - Select a printer
   - Select the paper orientation (Portrait or Landscape)
   - Set left, right, top, and bottom margins
3. Use the Headers/Footers page of the dialog box to:
   - Supply the text for a header that will appear on each page
   - Indicate whether the header should be printed or not
   - Supply the text for a footer that will appear on each page
   - Indicate whether the footer should be printed or not
   - Indicate whether or not pages should be numbered
4. Click **OK** to accept your choices.

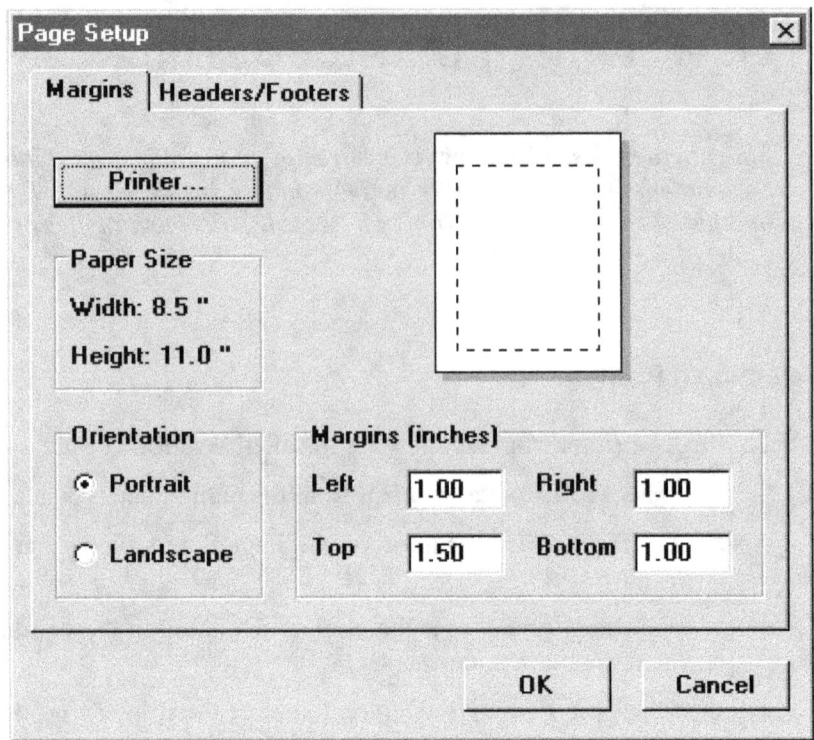

Figure 10.1 Page Setup Dialog

10.3 Print Preview

To preview a printout, select **File >> Print Preview** from the main menu. A Preview form will appear which shows how each page of the object being printed will appear when printed.

10.4 Printing the Current View

To print the contents of the current window being viewed in the EPANET workspace select **File >> Print** from the main menu or click ![print icon] on the Standard Toolbar. The following views can be printed:

- Data Browser (properties of the currently selected object)
- Network Map (at the current zoom level)
- Graphs (Time Series, Profile, Contour, Frequency and System Flow plots)
- Tables (Network and Time Series tables)
- Status, Energy, Calibration, and Reaction Reports.

10.5 Copying to the Clipboard or to a File

EPANET can copy the text and graphics of the current window being viewed to both the Windows clipboard and to a file. Views that can be copied in this fashion include the Network Map, graphs, tables, and reports. To copy the current view to the clipboard or to file:

1. Select **Edit >> Copy To** from the main menu or click ![icon].
2. Select choices from the Copy dialog that appears (see Figure 10.2) and click its **OK** button.
3. If you selected to copy-to-file, enter the name of the file in the Save As dialog box that appears and click **OK**.

Use the Copy dialog as follows to define how you want your data copied and to where:

1. Select a destination for the material being copied (Clipboard or File)
2. Select a format to copy in:
   - Bitmap (graphics only)
   - Metafile (graphics only)
   - Data (text, selected cells in a table, or data used to construct a graph)
3. Click **OK** to accept your selections or **Cancel** to cancel the copy request.

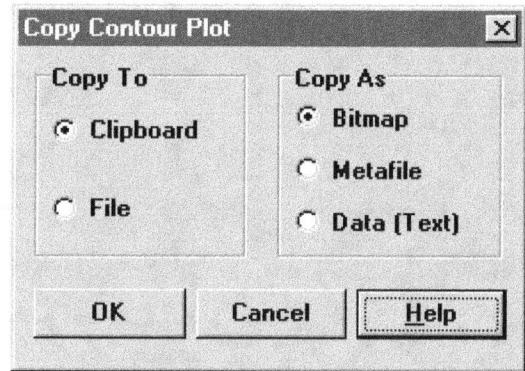

Figure 10.2 Copy Dialog

(This page intentionally left blank.)

CHAPTER 11 - IMPORTING AND EXPORTING

This chapter introduces the concept of Project Scenarios and describes how EPANET can import and export these and other data, such as the network map and the entire project database.

11.1 Project Scenarios

A Project Scenario consists of a subset of the data that characterizes the current conditions under which a pipe network is being analyzed. A scenario can consist of one or more of the following data categories:

- Demands (baseline demand plus time patterns for all categories) at all nodes
- Initial water quality at all nodes
- Diameters for all pipes
- Roughness coefficients for all pipes
- Reaction coefficients (bulk and wall) for all pipes
- Simple and rule-based controls

EPANET can compile a scenario based on some or all of the data categories listed above, save the scenario to file, and read the scenario back in at a later time.

Scenarios can provide more efficient and systematic analysis of design and operating alternatives. They can be used to examine the impacts of different loading conditions, search for optimal parameter estimates, and evaluate changes in operating policies. The scenario files are saved as ASCII text and can be created or modified outside of EPANET using a text editor or spreadsheet program.

11.2 Exporting a Scenario

To export a project scenario to a text file:

1. Select **File >> Export >> Scenario** from the main menu.
2. In the Export Data dialog form that appears (see Figure 11.1) select the types of data that you wish to save.
3. Enter an optional description of the scenario you are saving in the Notes memo field.
4. Select the **OK** button to accept your choices.
5. In the Save dialog box that next appears select a folder and name for the scenario file. Scenario files use the default extension .SCN.
6. Click **OK** to complete the export.

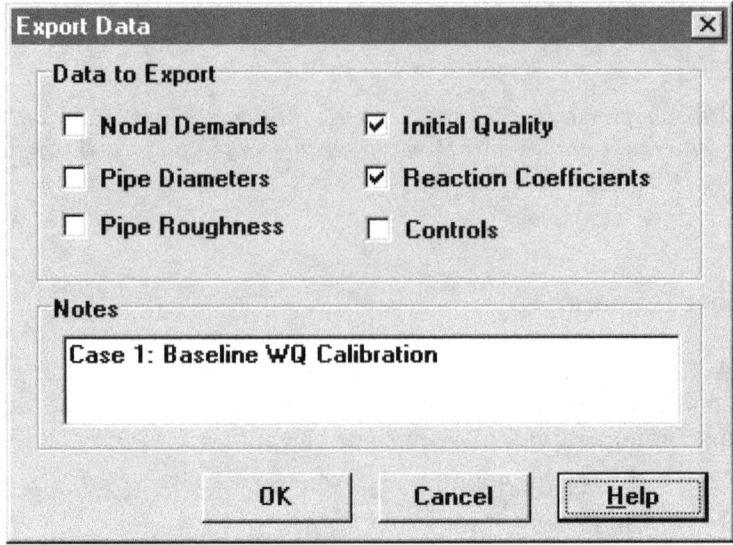

Figure 11.1 Export Data Dialog

The exported scenario can be imported back into the project at a later time as described in the next section.

11.3 Importing a Scenario

To import a project scenario from a file:

1. Select **File >> Import >> Scenario** from the main menu.

2. Use the Open File dialog box that appears to select a scenario file to import. The dialog's Contents panel will display the first several lines of files as they are selected, to help locate the desired file.

3. Click the **OK** button to accept your selection.

The data contained in the scenario file will replace any existing of the same kind in the current project.

11.4 Importing a Partial Network

EPANET has the ability to import a geometric description of a pipe network in a simple text format. This description simply contains the ID labels and map coordinates of the nodes and the ID labels and end nodes of the links. This simplifies the process of using other programs, such as CAD and GIS packages, to digitize network geometric data and then transfer these data to EPANET.

The format of a partial network text file looks as follows, where the text between brackets (< >) describes what type of information appears in that line of the file:

```
[TITLE]
```
<optional description of the file>

```
[JUNCTIONS]
```
<ID label of each junction>

```
[PIPES]
```
<ID label of each pipe followed by the ID labels of its end junctions>

```
[COORDINATES]
```
<Junction ID and its X and Y coordinates>

```
[VERTICES]
```
<Pipe ID and the X and Y coordinates of an intermediate vertex point>

Note that only junctions and pipes are represented. Other network elements, such as reservoirs and pumps, can either be imported as junctions or pipes and converted later on or simply be added in later on. The user is responsible for transferring any data generated from a CAD or GIS package into a text file with the format shown above.

In addition to this partial representation, a complete specification of the network can be placed in a file using the format described in Appendix C. This is the same format EPANET uses when a project is exported to a text file (see Section 11.7 below). In this case the file would also contain information on node and link properties, such as elevations, demands, diameters, roughness, etc.

11.5 Importing a Network Map

To import the coordinates for a network map stored in a text file:

1. Select **File >> Import >> Map** from the main menu.
2. Select the file containing the map information from the Open File dialog that appears.
3. Click **OK** to replace the current network map with the one described in the file.

11.6 Exporting the Network Map

The current view of the network map can be saved to file using either Autodesk's DXF (Drawing Exchange Format) format, the Windows enhanced metafile (EMF) format, or EPANET's own ASCII text (map) format. The DXF format is readable by many Computer Aided Design (CAD) programs. Metafiles can be inserted into word processing documents and loaded into drawing programs for re-scaling and editing. Both formats are vector-based and will not loose resolution when they are displayed at different scales.

To export the network map at full extent to a DXF, metafile, or text file:

1. Select **File >> Export >> Map** from the main menu.

2. In the Map Export dialog form that appears (see Figure 11.2) select the format that you want the map saved in.

3. If you select DXF format, you have a choice of how junctions will be represented in the DXF file. They can be drawn as open circles, as filled circles, or as filled squares. Not all DXF readers can recognize the commands used in the DXF file to draw a filled circle.

4. After choosing a format, click OK and enter a name for the file in the Save As dialog form that appears.

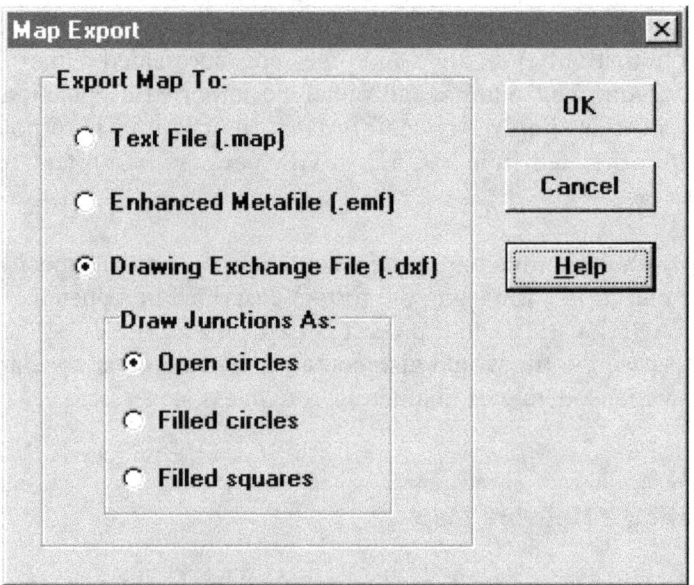

Figure 11.2 Map Export Dialog

11.7 Exporting to a Text File

To export a project's data to a text file:

1. Select **File >> Export >> Network** from the main menu.

2. In the Save dialog form that appears enter a name for the file to save to (the default extension is .INP).

3. Click **OK** to complete the export.

The resulting file will be written in ASCII text format, with the various data categories and property labels clearly identified. It can be read back into EPANET for analysis at another time by using either the **File >> Open** or **File >> Import >> Network** commands. Complete network descriptions using this input format can also be created outside of EPANET using any text editor or spreadsheet program. A complete specification of the .INP file format is given in Appendix C.

It is a good idea to save an archive version of your database in this format so you have access to a human readable version of your data. However, for day-to-day use of EPANET it is more efficient to save your data using EPANET's special project file format (that creates a .NET file) by using the **File >> Save** or **File >> Save As** commands. This format contains additional project information, such as the colors and ranges chosen for the map legends, the set of map display options in effect, the names of registered calibration data files, and any printing options that were selected.

(This page intentionally left blank.)

CHAPTER 12 - FREQUENTLY ASKED QUESTIONS

How can I import a pipe network created with a CAD or GIS program?

See Section 11.4.

How do I model a groundwater pumping well?

Represent the well as a reservoir whose head equals the piezometric head of the groundwater aquifer. Then connect your pump from the reservoir to the rest of the network. You can add piping ahead of the pump to represent local losses around the pump.

If you know the rate at which the well is pumping then an alternate approach is to replace the well – pump combination with a junction assigned a negative demand equal to the pumping rate. A time pattern can also be assigned to the demand if the pumping rate varies over time.

How do I size a pump to meet a specific flow?

Set the status of the pump to CLOSED. At the suction (inlet) node of the pump add a demand equal to the required pump flow and place a negative demand of the same magnitude at the discharge node. After analyzing the network, the difference in heads between the two nodes is what the pump needs to deliver.

How do I size a pump to meet a specific head?

Replace the pump with a Pressure Breaker Valve oriented in the opposite direction. Convert the design head to an equivalent pressure and use this as the setting for the valve. After running the analysis the flow through the valve becomes the pump's design flow.

How can I enforce a specific schedule of source flows into the network from my reservoirs?

Replace the reservoirs with junctions that have negative demands equal to the schedule of source flows. (Make sure there is at least one tank or remaining reservoir in the network, otherwise EPANET will issue an error message.)

How can I analyze fire flow conditions for a particular junction node?

To determine the maximum pressure available at a node when the flow demanded must be increased to suppress a fire, add the fire flow to the node's normal demand, run the analysis, and note the resulting pressure at the node.

To determine the maximum flow available at a particular pressure, set the emitter coefficient at the node to a large value (e.g., 100 times the maximum expected flow) and add the required pressure head (2.3 times the pressure in psi) to the node's elevation. After running the analysis, the available fire flow equals the actual demand reported for the node minus any consumer demand that was assigned to it.

How do I model a reduced pressure backflow prevention valve?

Use a General Purpose Valve with a headloss curve that shows increasing head loss with decreasing flow. Information from the valve manufacturer should provide help in constructing the curve. Place a check valve (i.e., a short length of pipe whose status is set to CV) in series with the valve to restrict the direction of flow.

How do I model a pressurized pneumatic tank?

If the pressure variation in the tank is negligible, use a very short, very wide cylindrical tank whose elevation is set close to the pressure head rating of the tank. Select the tank dimensions so that changes in volume produce only very small changes in water surface elevation.

If the pressure head developed in the tank ranges between H1 and H2, with corresponding volumes V1 and V2, then use a cylindrical tank whose cross-sectional area equals (V2-V1)/(H2-H1).

How do I model a tank inlet that discharges above the water surface?

Use the configuration shown below:

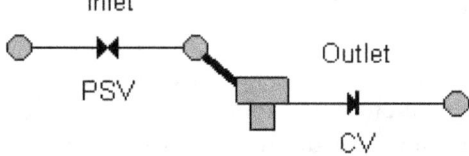

The tank's inlet consists of a Pressure Sustaining Valve followed by a short length of large diameter pipe. The pressure setting of the PSV should be 0, and the elevation of its end nodes should equal the elevation at which the true pipe connects to the tank. Use a Check Valve on the tank's outlet line to prevent reverse flow through it.

How do I determine initial conditions for a water quality analysis?

If simulating existing conditions monitored as part of a calibration study, assign measured values to the nodes where measurements were made and interpolate (by eye) to assign values to other locations. It is highly recommended that storage tanks and source locations be included in the set of locations where measurements are made.

To simulate future conditions start with arbitrary initial values (except at the tanks) and run the analysis for a number of repeating demand pattern cycles so that the water quality results begin to repeat in a periodic fashion as well. The number of such cycles can be reduced if good initial estimates are made for the water quality in the tanks. For example, if modeling water age the initial value could be set to the tank's average residence time, which is approximately equal to the fraction of its volume it exchanges each day.

How do I estimate values of the bulk and wall reaction coefficients?

Bulk reaction coefficients can be estimated by performing a bottle test in the laboratory (see Bulk Reactions in Section 3.4). Wall reaction rates cannot be measured directly. They must be back-fitted against calibration data collected from field studies (e.g., using trial and error to determine coefficient values that produce simulation results that best match field observations). Plastic pipe and relatively new lined iron pipe are not expected to exert any significant wall demand for disinfectants such as chlorine and chloramines.

How can I model a chlorine booster station?

Place the booster station at a junction node with zero or positive demand or at a tank. Select the node into the Property Editor and click the ellipsis button in the Source Quality field to launch the Source Quality Editor. In the editor, set Source Type to SETPOINT BOOSTER and set Source Quality to the chlorine concentration that water leaving the node will be boosted to. Alternatively, if the booster station will use flow-paced addition of chlorine then set Source Type to FLOW PACED BOOSTER and Source Quality to the concentration that will be added to the concentration leaving the node. Specify a time pattern ID in the Time Pattern field if you wish to vary the boosting level with time.

How would I model THM growth in a network?

THM growth can be modeled using first-order saturation kinetics. Select Options – Reactions from the Data Browser. Set the bulk reaction order to 1 and the limiting concentration to the maximum THM level that the water can produce, given a long enough holding time. Set the bulk reaction coefficient to a positive number reflective of the rate of THM production (e.g., 0.7 divided by the THM doubling time). Estimates of the reaction coefficient and the limiting concentration can be obtained from laboratory testing. The reaction coefficient will increase with increasing water temperature. Initial concentrations at all network nodes should at least equal the THM concentration entering the network from its source node.

Can I use a text editor to edit network properties while running EPANET?

Save the network to file as ASCII text (select **File >> Export >> Network**). With EPANET still running, start up your text editor program. Load the saved network file into the editor. When you are done editing the file, save it to disk. Switch to EPANET and read in the file (select **File >> Open**). You can keep switching back and forth between the editor program and EPANET, as more changes are needed. Just remember to save the file after modifying it in the editor, and re-open it again after switching to EPANET. If you use a word processor (such as WordPad) or a spreadsheet as your editor, remember to save the file as plain ASCII text.

Can I run multiple EPANET sessions at the same time?

Yes. This could prove useful in making side-by-side comparisons of two or more different design or operating scenarios.

APPENDIX A - UNITS OF MEASUREMENT

PARAMETER	US CUSTOMARY	SI METRIC
Concentration	mg/L or µg/L	mg/L or µg/L
Demand	(see Flow units)	(see Flow units)
Diameter (Pipes)	inches	millimeters
Diameter (Tanks)	feet	meters
Efficiency	percent	percent
Elevation	feet	meters
Emitter Coefficient	flow units / (psi)$^{1/2}$	flow units / (meters)$^{1/2}$
Energy	kilowatt - hours	kilowatt - hours
Flow	CFS (cubic feet / sec) GPM (gallons / min) MGD (million gal / day) IMGD (Imperial MGD) AFD (acre-feet / day)	LPS (liters / sec) LPM (liters / min) MLD (megaliters / day) CMH (cubic meters / hr) CMD (cubic meters / day)
Friction Factor	unitless	unitless
Hydraulic Head	feet	meters
Length	feet	meters
Minor Loss Coeff.	unitless	unitless
Power	horsepower	kilowatts
Pressure	pounds per square inch	meters
Reaction Coeff. (Bulk)	1/day (1st-order)	1/day (1st-order)
Reaction Coeff. (Wall)	mass / L / day (0-order) ft / day (1st-order)	mass / L / day (0-order) meters / day (1st-order)
Roughness Coefficient	10^{-3} feet (Darcy-Weisbach), unitless otherwise	millimeters (Darcy-Weisbach), unitless otherwise
Source Mass Injection	mass / minute	mass / minute
Velocity	feet / second	meters / second
Volume	cubic feet	cubic meters
Water Age	hours	hours

Note: US Customary units apply when CFS, GPM, AFD, or MGD is chosen as flow units. SI Metric units apply when flow units are expressed using either liters or cubic meters.

(This page intentionally left blank.)

APPENDIX B - ERROR MESSAGES

ID	Explanation
101	An analysis was terminated due to insufficient memory available.
110	An analysis was terminated because the network hydraulic equations could not be solved. Check for portions of the network not having any physical links back to a tank or reservoir or for unreasonable values for network input data.
200	One or more errors were detected in the input data. The nature of the error will be described by the 200-series error messages listed below.
201	There is a syntax error in a line of the input file created from your network data. This is most likely to have occurred in .INP text created by a user outside of EPANET.
202	An illegal numeric value was assigned to a property.
203	An object refers to undefined node.
204	An object refers to an undefined link.
205	An object refers to an undefined time pattern.
206	An object refers to an undefined curve.
207	An attempt is made to control a check valve. Once a pipe is assigned a Check Valve status with the Property Editor, its status cannot be changed by either simple or rule-based controls.
208	Reference was made to an undefined node. This could occur in a control statement for example.
209	An illegal value was assigned to a node property.
210	Reference was made to an undefined link. This could occur in a control statement for example.
211	An illegal value was assigned to a link property.
212	A source tracing analysis refers to an undefined trace node.
213	An analysis option has an illegal value (an example would be a negative time step value).
214	There are too many characters in a line read from an input file. The lines in the .INP file are limited to 255 characters.
215	Two or more nodes or links share the same ID label.
216	Energy data were supplied for an undefined pump.
217	Invalid energy data were supplied for a pump.
219	A valve is illegally connected to a reservoir or tank. A PRV, PSV or FCV cannot be directly connected to a reservoir or tank. Use a length of pipe to separate the two.

220	A valve is illegally connected to another valve. PRVs cannot share the same downstream node or be linked in series, PSVs cannot share the same upstream node or be linked in series, and a PSV cannot be directly connected to the downstream node of a PRV.
221	A rule-based control contains a misplaced clause.
223	There are not enough nodes in the network to analyze. A valid network must contain at least one tank/reservoir and one junction node.
224	There is not at least one tank or reservoir in the network.
225	Invalid lower/upper levels were specified for a tank (e.g., the lower lever is higher than the upper level).
226	No pump curve or power rating was supplied for a pump. A pump must either be assigned a curve ID in its Pump Curve property or a power rating in its Power property. If both properties are assigned then the Pump Curve is used.
227	A pump has an invalid pump curve. A valid pump curve must have decreasing head with increasing flow.
230	A curve has non-increasing X-values.
233	A node is not connected to any links.
302	The system cannot open the temporary input file. Make sure that the EPANET Temporary Folder selected has write privileges assigned to it (see Section 4.9).
303	The system cannot open the status report file. See Error 302.
304	The system cannot open the binary output file. See Error 302.
308	Could not save results to file. This can occur if the disk becomes full.
309	Could not write results to report file. This can occur if the disk becomes full.

APPENDIX C - COMMAND LINE EPANET

C.1 General Instructions

EPANET can also be run as a console application from the command line within a DOS window. In this case network input data are placed into a text file and results are written to a text file. The command line for running EPANET in this fashion is:

```
epanet2d  inpfile  rptfile  outfile
```

Here **inpfile** is the name of the input file, **rptfile** is the name of the output report file, and **outfile** is the name of an optional binary output file that stores results in a special binary format. If the latter file is not needed then just the input and report file names should be supplied. As written, the above command assumes that you are working in the directory in which EPANET was installed or that this directory has been added to the PATH statement in your AUTOEXEC.BAT file. Otherwise full pathnames for the executable **epanet2d.exe** and the files on the command line must be used. The error messages for command line EPANET are the same as those for Windows EPANET and are listed in Appendix B.

C.2 Input File Format

The input file for command line EPANET has the same format as the text file that Windows EPANET generates from its **File >> Export >> Network** command. It is organized in sections, where each section begins with a keyword enclosed in brackets. The various keywords are listed below.

Network Components	System Operation	Water Quality	Options and Reporting	Network Map/Tags
[TITLE]	[CURVES]	[QUALITY]	[OPTIONS]	[COORDINATES]
[JUNCTIONS]	[PATTERNS]	[REACTIONS]	[TIMES]	[VERTICES]
[RESERVOIRS]	[ENERGY]	[SOURCES]	[REPORT]	[LABELS]
[TANKS]	[STATUS]	[MIXING]		[BACKDROP]
[PIPES]	[CONTROLS]			[TAGS]
[PUMPS]	[RULES]			
[VALVES]	[DEMANDS]			
[EMITTERS]				

The order of sections is not important. However, whenever a node or link is referred to in a section it must have already been defined in the [JUNCTIONS], [RESERVOIRS], [TANKS], [PIPES], [PUMPS], or [VALVES] sections. Thus it is recommended that these sections be placed first, right after the [TITLE] section. The network map and tags sections are not used by command line EPANET and can be eliminated from the file.

Each section can contain one or more lines of data. Blank lines can appear anywhere in the file and the semicolon (;) can be used to indicate that what follows on the line is a comment, not data. A maximum of 255 characters can appear on a line. The ID labels used to identify nodes, links, curves and patterns can be any combination of up to 15 characters and numbers.

Figure C.1 displays the input file that represents the tutorial network discussed in Chapter 2.

```
[TITLE]
EPANET TUTORIAL

[JUNCTIONS]
;ID    Elev    Demand
;------------------
2      0       0
3      710     650
4      700     150
5      695     200
6      700     150

[RESERVOIRS]
;ID    Head
;---------
1      700

[TANKS]
;ID    Elev    InitLvl   MinLvl   MaxLvl   Diam   Volume
;-------------------------------------------------------
7      850     5         0        15       70     0

[PIPES]
;ID    Node1   Node2   Length   Diam   Roughness
;-----------------------------------------------
1      2       3       3000     12     100
2      3       6       5000     12     100
3      3       4       5000     8      100
4      4       5       5000     8      100
5      5       6       5000     8      100
6      6       7       7000     10     100

[PUMPS]
;ID    Node1   Node2   Parameters
;--------------------------------
7      1       2       HEAD 1
```

Figure C.1 Example EPANET Input File (continued on next page)

```
[PATTERNS]
;ID    Multipliers
;-----------------------
1      0.5  1.3  1  1.2
[CURVES]
;ID  X-Value   Y-Value
;---------------------
1    1000      200

[QUALITY]
;Node InitQual
;-------------
1     1

[REACTIONS]
Global Bulk   -1
Global Wall   0

[TIMES]
Duration               24:00
Hydraulic Timestep     1:00
Quality Timestep       0:05
Pattern Timestep       6:00

[REPORT]
Page      55
Energy    Yes
Nodes     All
Links     All

[OPTIONS]
Units              GPM
Headloss           H-W
Pattern            1
Quality            Chlorine mg/L
Tolerance          0.01

[END]
```

Figure C.1 Example EPANET Input File (continued from previous page)

On the pages that follow the contents and formats of each keyword section are described in alphabetical order.

[BACKDROP]

Purpose:

Identifies a backdrop image and dimensions for the network map.

Formats:

```
DIMENSIONS      LLx  LLy  URx  URy
UNITS           FEET/METERS/DEGREES/NONE
FILE            filename
OFFSET          X  Y
```

Definitions:

> `DIMENSIONS` provides the X and Y coordinates of the lower-left and upper-right corners of the map's bounding rectangle. Defaults are the extents of the nodal coordinates supplied in the [COORDINATES] section.
>
> `UNITS` specifies the units that the map's dimensions are given in. Default is NONE.
>
> `FILE` is the name of the file that contains the backdrop image.
>
> `OFFSET` lists the X and Y distance that the upper-left corner of the backdrop image is offset from the upper-left corner of the map's bounding rectangle. Default is zero offset.

Remarks:

a. The [BACKDROP] section is optional and is not used at all when EPANET is run as a console application.
b. Only Windows Enhanced Metafiles and bitmap files can be used as backdrops.

[CONTROLS]

Purpose:

Defines simple controls that modify links based on a single condition.

Format:

One line for each control which can be of the form:

    LINK linkID status **IF NODE** nodeID **ABOVE/BELOW** value

    LINK linkID status **AT TIME** time

    LINK linkID status **AT CLOCKTIME** clocktime **AM/PM**

where:

linkID	=	a link ID label
status	=	OPEN or CLOSED, a pump speed setting, or a control valve setting
nodeID	=	a node ID label
value	=	a pressure for a junction or a water level for a tank
time	=	a time since the start of the simulation in decimal hours or in hours:minutes format
clocktime	=	a 24-hour clock time (hours:minutes)

Remarks:

a. Simple controls are used to change link status or settings based on tank water level, junction pressure, time into the simulation or time of day.

b. See the notes for the [STATUS] section for conventions used in specifying link status and setting, particularly for control valves.

Examples:

```
[CONTROLS]
;Close Link 12 if the level in Tank 23 exceeds 20 ft.
LINK 12 CLOSED IF NODE 23 ABOVE 20

;Open Link 12 if pressure at Node 130 is under 30 psi
LINK 12 OPEN IF NODE 130 BELOW 30

;Pump PUMP02's speed is set to 1.5 at 16 hours into
;the simulation
LINK PUMP02 1.5 AT TIME 16

;Link 12 is closed at 10 am and opened at 8 pm
;throughout the simulation
LINK 12 CLOSED AT CLOCKTIME 10 AM
LINK 12 OPEN AT CLOCKTIME 8 PM
```

[COORDINATES]

Purpose:

Assigns map coordinates to network nodes.

Format:

One line for each node containing:

- Node ID label
- X-coordinate
- Y-coordinate

Remarks:

a. Include one line for each node displayed on the map.
b. The coordinates represent the distance from the node to an arbitrary origin at the lower left of the map. Any convenient units of measure for this distance can be used.
c. There is no requirement that all nodes be included in the map, and their locations need not be to actual scale.
d. A [COORDINATES] section is optional and is not used at all when EPANET is run as a console application.

Example:

```
[COORDINATES]
;Node         X-Coord.     Y-Coord
;------------------------------
   1          10023        128
   2          10056        95
```

[CURVES]

Purpose:

Defines data curves and their X,Y points.

Format:

One line for each X,Y point on each curve containing:

- Curve ID label
- X value
- Y value

Remarks:

a. Curves can be used to represent the following relations:
   - Head v. Flow for pumps
   - Efficiency v. Flow for pumps
   - Volume v. Depth for tanks
   - Headloss v. Flow for General Purpose Valves

b. The points of a curve must be entered in order of increasing X-values (lower to higher).

c. If the input file will be used with the Windows version of EPANET, then adding a comment which contains the curve type and description, separated by a colon, directly above the first entry for a curve will ensure that these items appear correctly in EPANET's Curve Editor. Curve types include PUMP, EFFICIENCY, VOLUME, and HEADLOSS. See the examples below.

Example:

```
[CURVES]
;ID     Flow    Head
;PUMP:  Curve for Pump 1
C1      0       200
C1      1000    100
C1      3000    0

;ID     Flow    Effic.
;EFFICIENCY:
E1      200     50
E1      1000    85
E1      2000    75
E1      3000    65
```

[DEMANDS]

Purpose:

Supplement to [JUNCTIONS] section for defining multiple water demands at junction nodes.

Format:

One line for each category of demand at a junction containing:

- Junction ID label
- Base demand (flow units)
- Demand pattern ID (optional)
- Name of demand category preceded by a semicolon (optional)

Remarks:

a. Only use for junctions whose demands need to be changed or supplemented from entries in [JUNCTIONS] section.

b. Data in this section replaces any demand entered in [JUNCTIONS] section for the same junction.

c. Unlimited number of demand categories can be entered per junction.

a. If no demand pattern is supplied then the junction demand follows the Default Demand Pattern specified in the [OPTIONS] section or Pattern 1 if no default pattern is specified. If the default pattern (or Pattern 1) does not exist, then the demand remains constant.

Example:

```
[DEMANDS]
;ID      Demand    Pattern     Category
;-----------------------------------------
J1       100       101         ;Domestic
J1       25        102         ;School
J256     50        101         ;Domestic
```

[EMITTERS]

Purpose:

Defines junctions modeled as emitters (sprinklers or orifices).

Format:

One line for each emitter containing:

- Junction ID label
- Flow coefficient, flow units at 1 psi (1 meter) pressure drop

Remarks:

a. Emitters are used to model flow through sprinkler heads or pipe leaks.

b. Flow out of the emitter equals the product of the flow coefficient and the junction pressure raised to a power.

c. The power can be specified using the EMITTER EXPONENT option in the [OPTIONS] section. The default power is 0.5, which normally applies to sprinklers and nozzles.

d. Actual demand reported in the program's results includes both the normal demand at the junction plus flow through the emitter.

e. An [EMITTERS] section is optional.

[ENERGY]

Purpose:

Defines parameters used to compute pumping energy and cost.

Formats:

```
GLOBAL            PRICE/PATTERN/EFFIC   value
PUMP    PumpID    PRICE/PATTERN/EFFIC   value
DEMAND CHARGE     value
```

Remarks:

c. Lines beginning with the keyword **GLOBAL** are used to set global default values of energy price, price pattern, and pumping efficiency for all pumps.

d. Lines beginning with the keyword **PUMP** are used to override global defaults for specific pumps.

e. Parameters are defined as follows:

- **PRICE** = average cost per kW-hour,
- **PATTERN** = ID label of time pattern describing how energy price varies with time,
- **EFFIC** = either a single percent efficiency for global setting or the ID label of an efficiency curve for a specific pump,
- **DEMAND CHARGE** = added cost per maximum kW usage during the simulation period.

f. The default global pump efficiency is 75% and the default global energy price is 0.

g. All entries in this section are optional. Items offset by slashes (/) indicate allowable choices.

Example:

```
[ENERGY]
GLOBAL PRICE       0.05    ;Sets global energy price
GLOBAL PATTERN     PAT1    ;and time-of-day pattern
PUMP   23  PRICE   0.10    ;Overrides price for Pump 23
PUMP   23  EFFIC   E23     ;Assigns effic. curve to Pump 23
```

[JUNCTIONS]

Purpose:

Defines junction nodes contained in the network.

Format:

One line for each junction containing:

- ID label
- Elevation, ft (m)
- Base demand flow (flow units) (optional)
- Demand pattern ID (optional)

Remarks:

b. A [JUNCTIONS] section with at least one junction is required.
c. If no demand pattern is supplied then the junction demand follows the Default Demand Pattern specified in the [OPTIONS] section or Pattern 1 if no default pattern is specified. If the default pattern (or Pattern 1) does not exist, then the demand remains constant.
d. Demands can also be entered in the [DEMANDS] section and include multiple demand categories per junction.

Example:

```
[JUNCTIONS]
;ID      Elev.    Demand    Pattern
;-----------------------------------
J1       100      50        Pat1
J2       120      10                  ;Uses default demand pattern
J3       115                          ;No demand at this junction
```

[LABELS]

Purpose:

Assigns coordinates to map labels.

Format:

One line for each label containing:

- X-coordinate
- Y-coordinate
- Text of label in double quotes
- ID label of an anchor node (optional)

Remarks:

a. Include one line for each label on the map.
b. The coordinates refer to the upper left corner of the label and are with respect to an arbitrary origin at the lower left of the map.
c. The optional anchor node anchors the label to the node when the map is re-scaled during zoom-in operations.
d. The [LABELS] section is optional and is not used at all when EPANET is run as a console application.

Example:

```
[LABELS]
;X-Coord.    Y-Coord.    Label            Anchor
;-----------------------------------------------
   1230       3459       "Pump 1"
   34.57      12.75      "North Tank"     T22
```

[MIXING]

Purpose:

Identifies the model that governs mixing within storage tanks.

Format:

One line per tank containing:

- Tank ID label
- Mixing model (MIXED, 2COMP, FIFO, or LIFO)
- Compartment volume (fraction)

Remarks:

a. Mixing models include:
    - Completely Mixed (MIXED)
    - Two-Compartment Mixing (2COMP)
    - Plug Flow (FIFO)
    - Stacked Plug Flow (LIFO)

b. The compartment volume parameter only applies to the two-compartment model and represents the fraction of the total tank volume devoted to the inlet/outlet compartment.

c. The [MIXING] section is optional. Tanks not described in this section are assumed to be completely mixed.

Example:

```
[MIXING]
;Tank      Model
;----------------------
T12        LIFO
T23        2COMP      0.2
```

[OPTIONS]

Purpose:

Defines various simulation options.

Formats:

UNITS	CFS/GPM/MGD/IMGD/AFD/
	LPS/LPM/MLD/CMH/CMD
HEADLOSS	H-W/D-W/C-M
HYDRAULICS	USE/SAVE filename
QUALITY	NONE/CHEMICAL/AGE/TRACE id
VISCOSITY	value
DIFFUSIVITY	value
SPECIFIC GRAVITY	value
TRIALS	value
ACCURACY	value
UNBALANCED	STOP/CONTINUE/CONTINUE n
PATTERN	id
DEMAND MULTIPLIER	value
EMITTER EXPONENT	value
TOLERANCE	value
MAP	filename

Definitions:

UNITS sets the units in which flow rates are expressed where:

CFS	=	cubic feet per second
GPM	=	gallons per minute
MGD	=	million gallons per day
IMGD	=	Imperial MGD
AFD	=	acre-feet per day
LPS	=	liters per second
LPM	=	liters per minute
MLD	=	million liters per day
CMH	=	cubic meters per hour
CMD	=	cubic meters per day

For CFS, GPM, MGD, IMGD, and AFD other input quantities are expressed in US Customary Units. If flow units are in liters or cubic meters then Metric Units must be used for all other input quantities as

152

well. (See Appendix A. Units of Measurement). The default flow units are **GPM**.

HEADLOSS selects a formula to use for computing head loss for flow through a pipe. The choices are the Hazen-Williams (**H-W**), Darcy-Weisbach (**D-W**), or Chezy-Manning (**C-M**) formulas. The default is **H-W**.

The **HYDRAULICS** option allows you to either **SAVE** the current hydraulics solution to a file or **USE** a previously saved hydraulics solution. This is useful when studying factors that only affect water quality behavior.

QUALITY selects the type of water quality analysis to perform. The choices are **NONE, CHEMICAL, AGE**, and **TRACE**. In place of **CHEMICAL** the actual name of the chemical can be used followed by its concentration units (e.g., **CHLORINE mg/L**). If **TRACE** is selected it must be followed by the ID label of the node being traced. The default selection is **NONE** (no water quality analysis).

VISCOSITY is the kinematic viscosity of the fluid being modeled relative to that of water at 20 deg. C (1.0 centistoke). The default value is 1.0.

DIFFUSIVITY is the molecular diffusivity of the chemical being analyzed relative to that of chlorine in water. The default value is 1.0. Diffusivity is only used when mass transfer limitations are considered in pipe wall reactions. A value of 0 will cause EPANET to ignore mass transfer limitations.

SPECIFIC GRAVITY is the ratio of the density of the fluid being modeled to that of water at 4 deg. C (unitless).

TRIALS are the maximum number of trials used to solve network hydraulics at each hydraulic time step of a simulation. The default is 40.

ACCURACY prescribes the convergence criterion that determines when a hydraulic solution has been reached. The trials end when the sum of all flow changes from the previous solution divided by the total flow in all links is less than this number. The default is 0.001.

UNBALANCED determines what happens if a hydraulic solution cannot be reached within the prescribed number of **TRIALS** at some hydraulic time step into the simulation. **"STOP"** will halt the entire analysis at that point. **"CONTINUE"** will continue the analysis with a warning message issued. **"CONTINUE n"** will continue the search for a solution for another "n" trials with the status of all links held fixed at their current settings. The simulation will be continued at this point with a message issued about whether convergence was achieved or not. The default choice is **"STOP"**.

PATTERN provides the ID label of a default demand pattern to be applied to all junctions where no demand pattern was specified. If no such pattern exists in the [PATTERNS] section then by default the pattern consists of a single multiplier equal to 1.0. If this option is not used, then the global default demand pattern has a label of "1".

The **DEMAND MULTIPLIER** is used to adjust the values of baseline demands for all junctions and all demand categories. For example, a value of 2 doubles all baseline demands, while a value of 0.5 would halve them. The default value is 1.0.

EMITTER EXPONENT specifies the power to which the pressure at a junction is raised when computing the flow issuing from an emitter. The default is 0.5.

MAP is used to supply the name of a file containing coordinates of the network's nodes so that a map of the network can be drawn. It is not used for any hydraulic or water quality computations.

TOLERANCE is the difference in water quality level below which one can say that one parcel of water is essentially the same as another. The default is 0.01 for all types of quality analyses (chemical, age (measured in hours), or source tracing (measured in percent)).

Remarks:

a. All options assume their default values if not explicitly specified in this section.

b. Items offset by slashes (/) indicate allowable choices.

Example:

```
[OPTIONS]
UNITS         CFS
HEADLOSS      D-W
QUALITY       TRACE   Tank23
UNBALANCED    CONTINUE   10
```

[PATTERNS]

Purpose:

Defines time patterns.

Format:

One or more lines for each pattern containing:

- Pattern ID label
- One or more multipliers

Remarks:

a. Multipliers define how some base quantity (e.g., demand) is adjusted for each time period.

a. All patterns share the same time period interval as defined in the [TIMES] section.

b. Each pattern can have a different number of time periods.

c. When the simulation time exceeds the pattern length the pattern wraps around to its first period.

d. Use as many lines as it takes to include all multipliers for each pattern.

Example:

```
[PATTERNS]
;Pattern P1
P1      1.1     1.4     0.9     0.7
P1      0.6     0.5     0.8     1.0
;Pattern P2
P2      1       1       1       1
P2      0       0       1
```

[PIPES]

Purpose:

Defines all pipe links contained in the network.

Format:

One line for each pipe containing:

- ID label of pipe
- ID of start node
- ID of end node
- Length, ft (m)
- Diameter, inches (mm)
- Roughness coefficient
- Minor loss coefficient
- Status (OPEN, CLOSED, or CV)

Remarks:

a. Roughness coefficient is unitless for the Hazen-Williams and Chezy-Manning head loss formulas and has units of millifeet (mm) for the Darcy-Weisbach formula. Choice of head loss formula is supplied in the [OPTIONS] section.

b. Setting status to CV means that the pipe contains a check valve restricting flow to one direction.

c. If minor loss coefficient is 0 and pipe is OPEN then these two items can be dropped form the input line.

Example:

```
[PIPES]
;ID     Node1   Node2   Length  Diam.   Roughness   Mloss   Status
;-----------------------------------------------------------------
 P1     J1      J2      1200    12      120         0.2     OPEN
 P2     J3      J2       600     6      110         0       CV
 P3     J1      J10     1000    12      120
```

[PUMPS]

Purpose:

Defines all pump links contained in the network.

Format:

One line for each pump containing:
- ID label of pump
- ID of start node
- ID of end node
- Keyword and Value (can be repeated)

Remarks:

a. Keywords consists of:
   - **POWER** – power value for constant energy pump, hp (kW)
   - **HEAD** - ID of curve that describes head versus flow for the pump
   - **SPEED** - relative speed setting (normal speed is 1.0, 0 means pump is off)
   - **PATTERN** - ID of time pattern that describes how speed setting varies with time

b. Either **POWER** or **HEAD** must be supplied for each pump. The other keywords are optional.

Example:

```
[PUMPS]
;ID      Node1    Node2    Properties
;----------------------------------------
Pump1    N12      N32      HEAD Curve1
Pump2    N121     N55      HEAD Curve1    SPEED 1.2
Pump3    N22      N23      POWER 100
```

[QUALITY]

Purpose:

Defines initial water quality at nodes.

Format:

One line per node containing:
- Node ID label
- Initial quality

Remarks:

a. Quality is assumed to be zero for nodes not listed.
b. Quality represents concentration for chemicals, hours for water age, or percent for source tracing.
c. The [QUALITY] section is optional.

[REACTIONS]

Purpose:

Defines parameters related to chemical reactions occurring in the network.

Formats:

    ORDER BULK/WALL/TANK   value
    GLOBAL BULK/WALL   value
    BULK/WALL/TANK   pipeID value
    LIMITING POTENTIAL   value
    ROUGHNESS CORRELATION   value

Definitions:

ORDER is used to set the order of reactions occurring in the bulk fluid, at the pipe wall, or in tanks, respectively. Values for wall reactions must be either 0 or 1. If not supplied the default reaction order is 1.0.

GLOBAL is used to set a global value for all bulk reaction coefficients (pipes and tanks) or for all pipe wall coefficients. The default value is zero.

BULK, WALL, and TANK are used to override the global reaction coefficients for specific pipes and tanks.

LIMITING POTENTIAL specifies that reaction rates are proportional to the difference between the current concentration and some limiting potential value.

ROUGHNESS CORRELATION will make all default pipe wall reaction coefficients be related to pipe roughness in the following manner:

Head Loss Equation	Roughness Correlation
Hazen-Williams	F / C
Darcy-Weisbach	F / log(e/D)
Chezy-Manning	F*n

where F = roughness correlation, C = Hazen-Williams C-factor, e = Darcy-Weisbach roughness, D = pipe diameter, and n = Chezy-Manning roughness coefficient. The default value computed this way can be overridden for any pipe by using the **WALL** format to supply a specific value for the pipe.

Remarks:

a. Remember to use positive numbers for growth reaction coefficients and negative numbers for decay coefficients.

b. The time units for all reaction coefficients are 1/days.

c. All entries in this section are optional. Items offset by slashes (/) indicate allowable choices.

Example:

```
[REACTIONS]
ORDER WALL    0     ;Wall reactions are zero-order
GLOBAL BULK  -0.5   ;Global bulk decay coeff.
GLOBAL WALL  -1.0   ;Global wall decay coeff.
WALL   P220  -0.5   ;Pipe-specific wall coeffs.
WALL   P244  -0.7
```

[REPORT]

Purpose:

Describes the contents of the output report produced from a simulation.

Formats:

PAGESIZE	value
FILE	filename
STATUS	YES/NO/FULL
SUMMARY	YES/NO
ENERGY	YES/NO
NODES	NONE/ALL/node1 node2 ...
LINKS	NONE/ALL/link1 link2 ...
parameter	YES/NO
parameter	BELOW/ABOVE/PRECISION value

Definitions:

PAGESIZE sets the number of lines written per page of the output report. The default is 0, meaning that no line limit per page is in effect.

FILE supplies the name of a file to which the output report will be written (ignored by the Windows version of EPANET).

STATUS determines whether a hydraulic status report should be generated. If YES is selected the report will identify all network components that change status during each time step of the simulation. If FULL is selected, then the status report will also include information from each trial of each hydraulic analysis. This level of detail is only useful for de-bugging networks that become hydraulically unbalanced. The default is NO.

SUMMARY determines whether a summary table of number of network components and key analysis options is generated. The default is YES.

ENERGY determines if a table reporting average energy usage and cost for each pump is provided. The default is NO.

NODES identifies which nodes will be reported on. You can either list individual node ID labels or use the keywords NONE or ALL. Additional NODES lines can be used to continue the list. The default is NONE.

LINKS identifies which links will be reported on. You can either list individual link ID labels or use the keywords NONE or ALL. Additional LINKS lines can be used to continue the list. The default is NONE.

The "parameter" reporting option is used to identify which quantities are reported on, how many decimal places are displayed, and what kind of filtering should be used to limit output reporting. Node parameters that can be reported on include:

- Elevation
- Demand
- Head
- Pressure
- Quality.

Link parameters include:

- Length
- Diameter
- Flow
- Velocity
- Headloss
- Position (same as status – open, active, closed)
- Setting (Roughness for pipes, speed for pumps, pressure/flow setting for valves)
- Reaction (reaction rate)
- F-Factor (friction factor).

The default quantities reported are Demand, Head, Pressure, and Quality for nodes and Flow, Velocity, and Headloss for links. The default precision is two decimal places.

Remarks:

a. All options assume their default values if not explicitly specified in this section.

b. Items offset by slashes (/) indicate allowable choices.

c. The default is to not report on any nodes or links, so a **NODES** or **LINKS** option must be supplied if you wish to report results for these items.

d. For the Windows version of EPANET, the only [REPORT] option recognized is **STATUS**. All others are ignored.

Example:

The following example reports on nodes N1, N2, N3, and N17 and all links with velocity above 3.0. The standard node parameters (Demand, Head, Pressure, and Quality) are reported on while only Flow, Velocity, and F-Factor (friction factor) are displayed for links.

```
[REPORT]
NODES N1 N2 N3 N17
LINKS ALL
FLOW YES
VELOCITY PRECISION 4
F-FACTOR PRECISION 4
VELOCITY ABOVE 3.0
```

[RESERVOIRS]

Purpose:

Defines all reservoir nodes contained in the network.

Format:

One line for each reservoir containing:

- ID label
- Head, ft (m)
- Head pattern ID (optional)

Remarks:

a. Head is the hydraulic head (elevation + pressure head) of water in the reservoir.

b. A head pattern can be used to make the reservoir head vary with time.

c. At least one reservoir or tank must be contained in the network.

Example:

```
[RESERVOIRS]
;ID     Head    Pattern
;---------------------
 R1     512             ;Head stays constant
 R2     120     Pat1    ;Head varies with time
```

[RULES]

Purpose:

Defines rule-based controls that modify links based on a combination of conditions.

Format:

Each rule is a series of statements of the form:

```
RULE  ruleID
IF    condition_1
AND   condition_2
OR    condition_3
AND   condition_4
etc.
THEN  action_1
AND   action_2
etc.
ELSE  action_3
AND   action_4
etc.
PRIORITY value
```

where:

`ruleID`	=	an ID label assigned to the rule
`conditon_n`	=	a condition clause
`action_n`	=	an action clause
`Priority`	=	a priority value (e.g., a number from 1 to 5)

Condition Clause Format:

A condition clause in a Rule-Based Control takes the form of:

```
object id attribute relation value
```

where

`object`	=	a category of network object
`id`	=	the object's ID label
`attribute`	=	an attribute or property of the object
`relation`	=	a relational operator
`value`	=	an attribute value

Some example conditional clauses are:

```
JUNCTION 23 PRESSURE > 20
TANK T200 FILLTIME BELOW 3.5
LINK 44 STATUS IS OPEN
SYSTEM DEMAND >= 1500
SYSTEM CLOCKTIME = 7:30 AM
```

The Object keyword can be any of the following:

NODE	**LINK**	**SYSTEM**
JUNCTION	**PIPE**	
RESERVOIR	**PUMP**	
TANK	**VALVE**	

When **SYSTEM** is used in a condition no ID is supplied.

The following attributes can be used with Node-type objects:

DEMAND

HEAD

PRESSURE

The following attributes can be used with Tanks:

LEVEL

FILLTIME (hours needed to fill a tank)

DRAINTIME (hours needed to empty a tank)

These attributes can be used with Link-Type objects:

FLOW

STATUS (**OPEN**, **CLOSED**, or **ACTIVE**)

SETTING (pump speed or valve setting)

The **SYSTEM** object can use the following attributes:

DEMAND (total system demand)

TIME (hours from the start of the simulation expressed either as a decimal number or in hours:minutes format)

CLOCKTIME (24-hour clock time with **AM** or **PM** appended)

Relation operators consist of the following:

=	IS
<>	NOT
<	BELOW
>	ABOVE
<=	>=

Action Clause Format:

An action clause in a Rule-Based Control takes the form of:

```
object id STATUS/SETTING IS value
```

where

object	=	**LINK**, **PIPE**, **PUMP**, or **VALVE** keyword
id	=	the object's ID label
value	=	a status condition (**OPEN** or **CLOSED**), pump speed setting, or valve setting

Some example action clauses are:

```
LINK 23 STATUS IS CLOSED
PUMP P100 SETTING IS 1.5
VALVE 123 SETTING IS 90
```

Remarks:

a. Only the **RULE, IF** and **THEN** portions of a rule are required; the other portions are optional.

b. When mixing **AND** and **OR** clauses, the **OR** operator has higher precedence than **AND**, i.e.,

```
IF A or B and C
```

is equivalent to

```
IF (A or B) and C.
```

If the interpretation was meant to be

```
IF A or (B and C)
```

then this can be expressed using two rules as in

```
IF A THEN ...
IF B and C THEN ...
```

c. The **PRIORITY** value is used to determine which rule applies when two or more rules require that conflicting actions be taken on a link. A rule without a priority value always has a lower priority than one with a value. For two rules with the same priority value, the rule that appears first is given the higher priority.

Example:

```
[RULES]
RULE 1
IF    TANK   1 LEVEL ABOVE 19.1
THEN  PUMP 335 STATUS IS CLOSED
AND   PIPE 330 STATUS IS OPEN

RULE 2
IF    SYSTEM CLOCKTIME >= 8 AM
AND   SYSTEM CLOCKTIME < 6 PM
AND   TANK 1 LEVEL BELOW 12
THEN  PUMP 335 STATUS IS OPEN

RULE 3
IF    SYSTEM CLOCKTIME >= 6 PM
OR    SYSTEM CLOCKTIME < 8 AM
AND   TANK 1 LEVEL BELOW 14
THEN  PUMP 335 STATUS IS OPEN
```

[SOURCES]

Purpose:

Defines locations of water quality sources.

Format:

One line for each water quality source containing:

- Node ID label
- Source type (CONCEN, MASS, FLOWPACED, or SETPOINT)
- Baseline source strength
- Time pattern ID (optional)

Remarks:

a. For **MASS** type sources, strength is measured in mass flow per minute. All other types measure source strength in concentration units.

b. Source strength can be made to vary over time by specifying a time pattern.

c. A **CONCEN** source:
   - represents the concentration of any external source inflow to the node
   - applies only when the node has a net negative demand (water enters the network at the node)
   - if the node is a junction, reported concentration is the result of mixing the source flow and inflow from the rest of the network
   - if the node is a reservoir, the reported concentration is the source concentration
   - if the node is a tank, the reported concentration is the internal concentration of the tank
   - is best used for nodes that represent source water supplies or treatment works (e.g., reservoirs or nodes assigned a negative demand)
   - should not be used at storage tanks with simultaneous inflow/outflow.

d. A **MASS, FLOWPACED,** or **SETPOINT** source:
   - represents a booster source, where the substance is injected directly into the network irregardless of what the demand at the node is
   - affects water leaving the node to the rest of the network in the following way:
     - a **MASS** booster adds a fixed mass flow to that resulting from inflow to the node
     - a **FLOWPACED** booster adds a fixed concentration to the resultant inflow concentration at the node
     - a **SETPOINT** booster fixes the concentration of any flow leaving the node (as long as the concentration resulting from the inflows is below the setpoint)
   - the reported concentration at a junction or reservoir booster source is the concentration that results after the boosting is applied; the reported concentration for a tank with a booster source is the internal concentration of the tank

- is best used to model direct injection of a tracer or disinfectant into the network or to model a contaminant intrusion.

e. A [SOURCES] section is not needed for simulating water age or source tracing.

Example:

```
[SOURCES]
;Node   Type     Strength   Pattern
;-------------------------------
  N1    CONCEN   1.2        Pat1      ;Concentration varies with time
  N44   MASS     12                   ;Constant mass injection
```

[STATUS]

Purpose:

Defines initial status of selected links at the start of a simulation.

Format:

One line per link being controlled containing:

- Link ID label
- Status or setting

Remarks:

a. Links not listed in this section have a default status of **OPEN** (for pipes and pumps) or **ACTIVE** (for valves).

b. The status value can be **OPEN** or **CLOSED**. For control valves (e.g., PRVs, FCVs, etc.) this means that the valve is either fully opened or closed, not active at its control setting.

c. The setting value can be a speed setting for pumps or valve setting for valves.

d. The initial status of pipes can also be set in the [PIPES] section.

e. Check valves cannot have their status be preset.

f. Use [CONTROLS] or [RULES] to change status or setting at some future point in the simulation.

g. If a **CLOSED** or **OPEN** control valve is to become **ACTIVE** again, then its pressure or flow setting must be specified in the control or rule that re-activates it.

Example:

```
[STATUS]
; Link    Status/Setting
;----------------------
  L22     CLOSED          ;Link L22 is closed
  P14     1.5             ;Speed for pump P14
  PRV1    OPEN            ;PRV1 forced open
                          ;(overrides normal operation)
```

[TAGS]

Purpose:

Associates category labels (tags) with specific nodes and links.

Format:

One line for each node and link with a tag containing

- the keyword NODE or LINK
- the node or link ID label
- the text of the tag label (with no spaces)

Remarks:

a. Tags can be useful for assigning nodes to different pressure zones or for classifying pipes by material or age.
b. If a node or link's tag is not identified in this section then it is assumed to be blank.
c. The [TAGS] section is optional and has no effect on the hydraulic or water quality calculations.

Example:

```
[TAGS]
;Object   ID      Tag
;-------------------------------
 NODE     1001    Zone_A
 NODE     1002    Zone_A
 NODE       45    Zone_B
 LINK      201    UNCI-1960
 LINK      202    PVC-1985
```

[TANKS]

Purpose:

Defines all tank nodes contained in the network.

Format:

One line for each tank containing:

- ID label
- Bottom elevation, ft (m)
- Initial water level, ft (m)
- Minimum water level, ft (m)
- Maximum water level, ft (m)
- Nominal diameter, ft (m)
- Minimum volume, cubic ft (cubic meters)
- Volume curve ID (optional)

Remarks:

a. Water surface elevation equals bottom elevation plus water level.
b. Non-cylindrical tanks can be modeled by specifying a curve of volume versus water depth in the [CURVES] section.
c. If a volume curve is supplied the diameter value can be any non-zero number
d. Minimum volume (tank volume at minimum water level) can be zero for a cylindrical tank or if a volume curve is supplied.
e. A network must contain at least one tank or reservoir.

Example:

```
[TANKS]
;ID    Elev.   InitLvl   MinLvl   MaxLvl   Diam   MinVol   VolCurve
;-------------------------------------------------------------------
;Cylindrical tank
T1     100     15        5        25       120    0
;Non-cylindrical tank with arbitrary diameter
T2     100     15        5        25       1      0        VC1
```

[TIMES]

Purpose:

Defines various time step parameters used in the simulation.

Formats:

DURATION	Value (units)
HYDRAULIC TIMESTEP	Value (units)
QUALITY TIMESTEP	Value (units)
RULE TIMESTEP	Value (units)
PATTERN TIMESTEP	Value (units)
PATTERN START	Value (units)
REPORT TIMESTEP	Value (units)
REPORT START	Value (units)
START CLOCKTIME	Value (AM/PM)
STATISTIC	NONE/AVERAGED/ MINIMUM/MAXIMUM RANGE

Definitions:

DURATION is the duration of the simulation. Use 0 to run a single period snapshot analysis. The default is 0.

HYDRAULIC TIMESTEP determines how often a new hydraulic state of the network is computed. If greater than either the PATTERN or REPORT time step it will be automatically reduced. The default is 1 hour.

QUALITY TIMESTEP is the time step used to track changes in water quality throughout the network. The default is 1/10 of the hydraulic time step.

RULE TIMESTEP is the time step used to check for changes in system status due to activation of rule-based controls between hydraulic time steps. The default is 1/10 of the hydraulic time step.

PATTERN TIMESTEP is the interval between time periods in all time patterns. The default is 1 hour.

PATTERN START is the time offset at which all patterns will start. For example, a value of 6 hours would start the simulation with each pattern in the time period that corresponds to hour 6. The default is 0.

REPORT TIMESTEP sets the time interval between which output results are reported. The default is 1 hour.

REPORT START is the length of time into the simulation at which output results begin to be reported. The default is 0.

START CLOCKTIME is the time of day (e.g., 3:00 PM) at which the simulation begins. The default is 12:00 AM midnight.

STATISTIC determines what kind of statistical post-processing should be done on the time series of simulation results generated. **AVERAGED** reports a set of time-averaged results, **MINIMUM** reports only the minimum values, **MAXIMUM** the maximum values, and **RANGE** reports the difference between the minimum and maximum values. **NONE** reports the full time series for all quantities for all nodes and links and is the default.

Remarks:

a. Units can be **SECONDS (SEC)**, **MINUTES (MIN)**, **HOURS**, or **DAYS**. The default is hours.
b. If units are not supplied, then time values can be entered as decimal hours or in hours:minutes notation.
c. All entries in the [TIMES] section are optional. Items offset by slashes (/) indicate allowable choices.

Example:

```
[TIMES]
DURATION            240 HOURS
QUALITY TIMESTEP    3 MIN
REPORT START        120
STATISTIC           AVERAGED
START CLOCKTIME     6:00 AM
```

[TITLE]

Purpose:

Attaches a descriptive title to the network being analyzed.

Format:

Any number of lines of text.

Remarks:

The [TITLE] section is optional.

[VALVES]

Purpose:

Defines all control valve links contained in the network.

Format:

One line for each valve containing:

- ID label of valve
- ID of start node
- ID of end node
- Diameter, inches (mm)
- Valve type
- Valve setting
- Minor loss coefficient

Remarks:

a. Valve types and settings include:

Valve Type	Setting
PRV (pressure reducing valve)	Pressure, psi (m)
PSV (pressure sustaining valve)	Pressure, psi (m)
PBV (pressure breaker valve)	Pressure, psi (m)
FCV (flow control valve)	Flow (flow units)
TCV (throttle control valve)	Loss Coefficient
GPV (general purpose valve)	ID of head loss curve

b. Shutoff valves and check valves are considered to be part of a pipe, not a separate control valve component (see [PIPES])

[VERTICES]

Purpose:

Assigns interior vertex points to network links.

Format:

One line for each point in each link containing such points that includes:

- Link ID label
- X-coordinate
- Y-coordinate

Remarks:

a. Vertex points allow links to be drawn as polylines instead of simple straight-lines between their end nodes.
b. The coordinates refer to the same coordinate system used for node and label coordinates.
c. A [VERTICES] section is optional and is not used at all when EPANET is run as a console application.

Example:

```
[COORDINATES]
;Node        X-Coord.      Y-Coord
;-------------------------------
   1           10023         128
   2           10056          95
```

C.3 Report File Format

Statements supplied to the [REPORT] section of the input file control the contents of the report file generated from a command-line run of EPANET. A portion of the report generated from the input file of Figure C.1 is shown in Figure C.2. In general a report can contain the following sections:

- Status Section
- Energy Section
- Nodes Section
- Links Section

Status Section

The Status Section of the output report lists the initial status of all reservoirs, tanks, pumps, valves, and closed pipes as well as any changes in the status of these components as they occur over time in an extended period simulation. The status of reservoirs and tanks indicates whether they are filling or emptying. The status of links indicates whether they are open or closed and includes the relative speed setting for pumps and the pressure/flow setting for control valves. To include a Status Section in the report use the command **STATUS YES** in the [REPORT] section of the input file.

Using **STATUS FULL** will also produce a full listing of the convergence results for all iterations of each hydraulic analysis made during a simulation. This listing will also show which components are changing status during the iterations. This level of detail is only useful when one is trying to debug a run that fails to converge because a component's status is cycling.

Energy Section

The Energy Section of the output report lists overall energy consumption and cost for each pump in the network. The items listed for each pump include:

- Percent Utilization (percent of the time that the pump is on-line)
- Average Efficiency
- Kilowatt-hours consumed per million gallons (or cubic meters) pumped
- Average Kilowatts consumed
- Peak Kilowatts used
- Average cost per day

Also listed is the total cost per day for pumping and the total demand charge (cost based on the peak energy usage) incurred. To include an Energy Section in the report the command **ENERGY YES** must appear in the [REPORT] section of the input file.

```
***************************************************************
*                        E P A N E T                          *
*              Hydraulic and Water Quality                    *
*               Analysis for Pipe Networks                    *
*                      Version 2.0                            *
***************************************************************

EPANET TUTORIAL

     Input Data File ................... tutorial.inp
     Number of Junctions................ 5
     Number of Reservoirs............... 1
     Number of Tanks ................... 1
     Number of Pipes ................... 6
     Number of Pumps ................... 1
     Number of Valves .................. 0
     Headloss Formula .................. Hazen-Williams
     Hydraulic Timestep ................ 1.00 hrs
     Hydraulic Accuracy ................ 0.001000
     Maximum Trials .................... 40
     Quality Analysis .................. Chlorine
     Water Quality Time Step ........... 5.00 min
     Water Quality Tolerance ........... 0.01 mg/L
     Specific Gravity .................. 1.00
     Relative Kinematic Viscosity ...... 1.00
     Relative Chemical Diffusivity ..... 1.00
     Demand Multiplier ................. 1.00
     Total Duration .................... 24.00 hrs
     Reporting Criteria:
        All Nodes
        All Links

Energy Usage:
---------------------------------------------------------------
         Usage   Avg.    Kw-hr     Avg.      Peak      Cost
Pump     Factor  Effic.  /Mgal     Kw        Kw        /day
---------------------------------------------------------------
7        100.00  75.00   746.34    51.34     51.59     0.00
---------------------------------------------------------------
                                   Demand Charge:      0.00
                                   Total Cost:         0.00
```

Figure C.2 Excerpt from a Report File (continued on next page)

```
Node Results at 0:00 hrs:
---------------------------------------------------------------
         Demand      Head   Pressure   Chlorine
Node        gpm        ft        psi       mg/L
---------------------------------------------------------------
2          0.00    893.37     387.10       0.00
3        325.00    879.78      73.56       0.00
4         75.00    874.43      75.58       0.00
5        100.00    872.69      76.99       0.00
6         75.00    872.71      74.84       0.00
1      -1048.52    700.00       0.00       1.00   Reservoir
7        473.52    855.00       2.17       0.00   Tank

Link Results at 0:00 hrs:
---------------------------------------------------------------
          Flow   Velocity   Headloss
Link       gpm        fps    /1000ft
---------------------------------------------------------------
1       1048.52       2.97       4.53
2        558.33       1.58       1.41
3        165.19       1.05       1.07
4         90.19       0.58       0.35
5         -9.81       0.06       0.01
6        473.52       1.93       2.53
7       1048.52       0.00    -193.37   Pump

Node Results at 1:00 hrs:
---------------------------------------------------------------
         Demand      Head   Pressure   Chlorine
Node        gpm        ft        psi       mg/L
---------------------------------------------------------------
2          0.00    893.92     387.34       1.00
3        325.00    880.42      73.84       0.99
4         75.00    875.12      75.88       0.00
5        100.00    873.40      77.30       0.00
6         75.00    873.43      75.15       0.00
1      -1044.60    700.00       0.00       1.00   Reservoir
7        469.60    855.99       2.59       0.00   Tank

Link Results at 1:00 hrs:
---------------------------------------------------------------
          Flow   Velocity   Headloss
Link       gpm        fps    /1000ft
---------------------------------------------------------------
1       1044.60       2.96       4.50
2        555.14       1.57       1.40
3        164.45       1.05       1.06
4         89.45       0.57       0.34
5        -10.55       0.07       0.01
6        469.60       1.92       2.49
7       1044.60       0.00    -193.92   Pump
```

Figure C.2 Excerpt from a Report File (continued from previous page)

Nodes Section

The Nodes Section of the output report lists simulation results for those nodes and parameters identified in the [REPORT] section of the input file. Results are listed for each reporting time step of an extended period simulation. The reporting time step is specified in the [TIMES] section of the input file. Results at intermediate times when certain hydraulic events occur, such as pumps turning on or off or tanks closing because they become empty or full, are not reported.

To have nodal results reported the [REPORT] section of the input file must contain the keyword **NODES** followed by a listing of the ID labels of the nodes to be included in the report. There can be several such **NODES** lines in the file. To report results for all nodes use the command **NODES ALL**.

The default set of reported quantities for nodes includes Demand, Head, Pressure, and Water Quality. You can specify how many decimal places to use when listing results for a parameter by using commands such as **PRESSURE PRECISION 3** in the input file (i.e., use 3 decimal places when reporting results for pressure). The default precision is 2 decimal places for all quantities. You can filter the report to list only the occurrences of values below or above a certain value by adding statements of the form **PRESSURE BELOW 20** to the input file.

Links Section

The Links Section of the output report lists simulation results for those links and parameters identified in the [REPORT] section of the input file. The reporting times follow the same convention as was described for nodes in the previous section.

As with nodes, to have any results for links reported you must include the keyword **LINKS** followed by a list of link ID labels in the [REPORT] section of the input file. Use the command **LINKS ALL** to report results for all links.

The default parameters reported on for links includes Flow, Velocity, and Headloss. Diameter, Length, Water Quality, Status, Setting, Reaction Rate, and Friction Factor can be added to these by using commands such as **DIAMETER YES** or **DIAMETER PRECISION 0**. The same conventions used with node parameters for specifying reporting precision and filters also applies to links.

C.4   Binary Output File Format

If a third file name is supplied to the command line that runs EPANET then the results for all parameters for all nodes and links for all reporting time periods will be saved to this file in a special binary format. This file can be used for special post-processing purposes. Data written to the file are 4-byte integers, 4-byte floats, or fixed-size strings whose size is a multiple of 4 bytes. This allows the file to be divided conveniently into 4-byte records. The file consists of four sections of the following sizes in bytes:

Section	Size in bytes
Prolog	852 + 20*Nnodes + 36*Nlinks + 8*Ntanks
Energy Use	28*Npumps + 4
Extended Period	(16*Nnodes + 32*Nlinks)*Nperiods
Epilog	28

where

Nnodes	=	number of nodes (junctions + reservoirs + tanks)
Nlinks	=	number of links (pipes + pumps + valves)
Ntanks	=	number of tanks and reservoirs
Npumps	=	number of pumps
Nperiods	=	number of reporting periods

and all of these counts are themselves written to the file's Prolog or Epilog sections.

Prolog Section

The prolog section of the binary Output File contains the following data:

Item	Type	Number of Bytes
Magic Number (= 516114521)	Integer	4
Version (= 200)	Integer	4
Number of Nodes (Junctions + Reservoirs + Tanks)	Integer	4
Number of Reservoirs & Tanks	Integer	4
Number of Links (Pipes + Pumps + Valves)	Integer	4
Number of Pumps	Integer	4
Number of Valves	Integer	4
Water Quality Option 0 = none 1 = chemical 2 = age 3 = source trace	Integer	4
Index of Node for Source Tracing	Integer	4
Flow Units Option 0 = cfs 1 = gpm 2 = mgd 3 = Imperial mgd 4 = acre-ft/day 5 = liters/second 6 = liters/minute 7 = megaliters/day 8 = cubic meters/hour 9 = cubic meters/day	Integer	4

Pressure Units Option 0 = psi 1 = meters 2 = kPa	Integer	4
Statistics Flag 0 = no statistical processing 1 = results are time-averaged 2 = only minimum values reported 3 = only maximum values reported 4 = only ranges reported	Integer	4
Reporting Start Time (seconds)	Integer	4
Reporting Time Step (seconds)	Integer	4
Simulation Duration (seconds)	Integer	4
Problem Title (1st line)	Char	80
Problem Title (2nd line)	Char	80
Problem Title (3rd line)	Char	80
Name of Input File	Char	260
Name of Report File	Char	260
Name of Chemical	Char	16
Chemical Concentration Units	Char	16
ID Label of Each Node	Char	16
ID Label of Each Link	Char	16
Index of Start Node of Each Link	Integer	4*Nlinks
Index of End Node of Each Link	Integer	4*Nlinks
Type Code of Each Link 0 = Pipe with CV 1 = Pipe 2 = Pump 3 = PRV 4 = PSV 5 = PBV 6 = FCV 7 = TCV 8 = GPV	Integer	4*Nlinks
Node Index of Each Tank	Integer	4*Ntanks
Cross-Sectional Area of Each Tank	Float	4*Ntanks
Elevation of Each Node	Float	4*Nnodes
Length of Each Link	Float	4*Nlinks
Diameter of Each Link	Float	4*Nlinks

There is a one-to-one correspondence between the order in which the ID labels for nodes and links are written to the file and the index numbers of these components. Also, reservoirs are distinguished from tanks by having their cross-sectional area set to zero.

Energy Use Section

The Energy Use section of the binary output file immediately follows the Prolog section. It contains the following data:

Item	Type	Number of Bytes
Repeated for each pump:		
▪ Pump Index in List of Links	Float	4
▪ Pump Utilization (%)	Float	4
▪ Average Efficiency (%)	Float	4
▪ Average Kwatts/Million Gallons (/Meter^3)	Float	4
▪ Average Kwatts	Float	4
▪ Peak Kwatts	Float	4
▪ Average Cost Per Day	Float	4
Overall Peak Energy Usage	Float	4

The statistics reported in this section refer to the period of time between the start of the output reporting period and the end of the simulation.

Extended Period Section

The Extended Period section of the binary Output File contains simulation results for each reporting period of an analysis (the reporting start time and time step are written to the Output File's Prolog section and the number of steps is written to the Epilog section). For each reporting period the following values are written to the file:

Item	Type	Size in Bytes
Demand at Each Node	Float	4*Nnodes
Hydraulic Head at Each Node	Float	4*Nnodes
Pressure at Each Node	Float	4*Nnodes
Water Quality at Each Node	Float	4*Nnodes
Flow in Each Link (negative for reverse flow)	Float	4*Nlinks
Velocity in Each Link	Float	4*Nlinks
Headloss per 1000 Units of Length for Each Link (Negative of head gain for pumps and total head loss for valves)	Float	4*Nlinks
Average Water Quality in Each Link	Float	4*Nlinks
Status Code for Each Link 0 = closed (max. head exceeded) 1 = temporarily closed 2 = closed 3 = open 4 = active (partially open) 5 = open (max. flow exceeded) 6 = open (flow setting not met) 7 = open (pressure setting not met)	Float	4*Nlinks
Setting for Each Link: Roughness Coefficient for Pipes Speed for Pumps Setting for Valves	Float	4*Nlinks
Reaction Rate for Each Link (mass/L/day)	Float	4*Nlinks
Friction Factor for Each Link	Float	4*Nlinks

Epilog Section

The Epilog section of the binary output file contains the following data:

Item	Type	Number of Bytes
Average bulk reaction rate (mass/hr)	Float	4
Average wall reaction rate (mass/hr)	Float	4
Average tank reaction rate (mass/hr)	Float	4
Average source inflow rate (mass/hr)	Float	4
Number of Reporting Periods	Integer	4

Warning Flag:
  0 = no warnings
  1 = warnings were generated
	Integer	4

Magic Number (= 516114521)	Integer	4

The mass units of the reaction rates both here and in the Extended Period output depend on the concentration units assigned to the chemical being modeled. The reaction rates listed in this section refer to the average of the rates seen in all pipes (or all tanks) over the entire reporting period of the simulation.

(This page intentionally left blank.)

APPENDIX D - ANALYSIS ALGORITHMS

D.1 Hydraulics

The method used in EPANET to solve the flow continuity and headloss equations that characterize the hydraulic state of the pipe network at a given point in time can be termed a hybrid node-loop approach. Todini and Pilati (1987) and later Salgado et al. (1988) chose to call it the "Gradient Method". Similar approaches have been described by Hamam and Brameller (1971) (the "Hybrid Method) and by Osiadacz (1987) (the "Newton Loop-Node Method"). The only difference between these methods is the way in which link flows are updated after a new trial solution for nodal heads has been found. Because Todini's approach is simpler, it was chosen for use in EPANET.

Assume we have a pipe network with N junction nodes and NF fixed grade nodes (tanks and reservoirs). Let the flow-headloss relation in a pipe between nodes i and j be given as:

$$H_i - H_j = h_{ij} = rQ_{ij}^n + mQ_{ij}^2 \qquad \text{D.1}$$

where H = nodal head, h = headloss, r = resistance coefficient, Q = flow rate, n = flow exponent, and m = minor loss coefficient. The value of the resistance coefficient will depend on which friction headloss formula is being used (see below). For pumps, the headloss (negative of the head gain) can be represented by a power law of the form

$$h_{ij} = -\omega^2 (h_0 - r\,(Q_{ij}/\omega)^n)$$

where h_0 is the shutoff head for the pump, ω is a relative speed setting, and r and n are the pump curve coefficients. The second set of equations that must be satisfied is flow continuity around all nodes:

$$\sum_j Q_{ij} - D_i = 0 \qquad \text{for i = 1,... N.} \qquad \text{D.2}$$

where D_i is the flow demand at node i and by convention, flow into a node is positive. For a set of known heads at the fixed grade nodes, we seek a solution for all heads H_i and flows Q_{ij} that satisfy Eqs. (D.1) and (D.2).

The Gradient solution method begins with an initial estimate of flows in each pipe that may not necessarily satisfy flow continuity. At each iteration of the method, new nodal heads are found by solving the matrix equation:

$$\mathbf{AH} = \mathbf{F} \qquad \text{D.3}$$

where **A** = an (NxN) Jacobian matrix, **H** = an (Nx1) vector of unknown nodal heads, and **F** = an (Nx1) vector of right hand side terms

The diagonal elements of the Jacobian matrix are:

$$A_{ii} = \sum_j p_{ij}$$

while the non-zero, off-diagonal terms are:

$$A_{ij} = -p_{ij}$$

where p_{ij} is the inverse derivative of the headloss in the link between nodes i and j with respect to flow. For pipes,

$$p_{ij} = \frac{1}{nr|Q_{ij}|^{n-1} + 2m|Q_{ij}|}$$

while for pumps

$$p_{ij} = \frac{1}{n\omega^2 r(Q_{ij}/\omega)^{n-1}}$$

Each right hand side term consists of the net flow imbalance at a node plus a flow correction factor:

$$F_i = \left(\sum_j Q_{ij} - D_i\right) + \sum_j y_{ij} + \sum_f p_{if} H_f$$

where the last term applies to any links connecting node i to a fixed grade node f and the flow correction factor y_{ij} is:

$$y_{ij} = p_{ij}\left(r|Q_{ij}|^n + m|Q_{ij}|^2\right)\text{sgn}(Q_{ij})$$

for pipes and

$$y_{ij} = -p_{ij}\omega^2\left(h_0 - r(Q_{ij}/\omega)^n\right)$$

for pumps, where sgn(x) is 1 if x > 0 and -1 otherwise. (Q_{ij} is always positive for pumps.)

After new heads are computed by solving Eq. (D.3), new flows are found from:

$$Q_{ij} = Q_{ij} - \left(y_{ij} - p_{ij}(H_i - H_j)\right) \qquad \text{D.4}$$

If the sum of absolute flow changes relative to the total flow in all links is larger than some tolerance (e.g., 0.001), then Eqs. (D.3) and (D.4) are solved once again. The flow update formula (D.4) always results in flow continuity around each node after the first iteration.

EPANET implements this method using the following steps:

1. The linear system of equations D.3 is solved using a sparse matrix method based on node re-ordering (George and Liu, 1981). After re-ordering the nodes to minimize the amount of fill-in for matrix A, a symbolic factorization is carried out so that only the non-zero elements of A need be stored and operated on in memory. For extended period simulation this re-ordering and factorization is only carried out once at the start of the analysis.

2. For the very first iteration, the flow in a pipe is chosen equal to the flow corresponding to a velocity of 1 ft/sec, while the flow through a pump equals the design flow specified for the pump. (All computations are made with head in feet and flow in cfs).

3. The resistance coefficient for a pipe (r) is computed as described in Table 3.1. For the Darcy-Weisbach headloss equation, the friction factor f is computed by different equations depending on the flow's Reynolds Number (Re):

Hagen – Poiseuille formula for Re < 2,000 (Bhave, 1991):

$$f = \frac{64}{\text{Re}}$$

Swamee and Jain approximation to the Colebrook - White equation for Re > 4,000 (Bhave, 1991):

$$f = \frac{0.25}{\left[Ln\left(\frac{\varepsilon}{3.7d} + \frac{5.74}{\text{Re}^{0.9}} \right) \right]^2}$$

Cubic Interpolation From Moody Diagram for 2,000 < Re < 4,000 (Dunlop, 1991):

$$f = (X1 + R(X2 + R(X3 + X4)))$$
$$R = \frac{\text{Re}}{2000}$$
$$X1 = 7FA - FB$$
$$X2 = 0.128 - 17FA + 2.5FB$$
$$X3 = -0.128 + 13FA - 2FB$$
$$X4 = R(0.032 - 3FA + 0.5FB)$$
$$FA = (Y3)^{-2}$$

$$FB = FA\left(2 - \frac{0.00514215}{(Y2)(Y3)}\right)$$

$$Y2 = \frac{\varepsilon}{3.7d} + \frac{5.74}{\text{Re}^{0.9}}$$

$$Y3 = -0.86859 Ln\left(\frac{\varepsilon}{3.7d} + \frac{5.74}{4000^{0.9}}\right)$$

where ε = pipe roughness and d = pipe diameter.

4. The minor loss coefficient based on velocity head (K) is converted to one based on flow (m) with the following relation:

$$m = \frac{0.02517 K}{d^4}$$

5. Emitters at junctions are modeled as a fictitious pipe between the junction and a fictitious reservoir. The pipe's headloss parameters are $n = (1/\gamma)$, $r = (1/C)^n$, and $m = 0$ where C is the emitter's discharge coefficient and γ is its pressure exponent. The head at the fictitious reservoir is the elevation of the junction. The computed flow through the fictitious pipe becomes the flow associated with the emitter.

6. Open valves are assigned an r-value by assuming the open valve acts as a smooth pipe (f = 0.02) whose length is twice the valve diameter. Closed links are assumed to obey a linear headloss relation with a large resistance factor, i.e., $h = 10^8 Q$, so that $p = 10^{-8}$ and $y = Q$. For links where $(r+m)Q < 10^{-7}$, $p = 10^7$ and $y = Q/n$.

7. Status checks on pumps, check valves (CVs), flow control valves, and pipes connected to full/empty tanks are made after every other iteration, up until the 10th iteration. After this, status checks are made only after convergence is achieved. Status checks on pressure control valves (PRVs and PSVs) are made after each iteration.

8. During status checks, pumps are closed if the head gain is greater than the shutoff head (to prevent reverse flow). Similarly, check valves are closed if the headloss through them is negative (see below). When these conditions are not present, the link is re-opened. A similar status check is made for links connected to empty/full tanks. Such links are closed if the difference in head across the link would cause an empty tank to drain or a full tank to fill. They are re-opened at the next status check if such conditions no longer hold.

9. Simply checking if $h < 0$ to determine if a check valve should be closed or open was found to cause cycling between these two states in some networks due to limits on numerical precision. The following procedure was devised to provide a more robust test of the status of a check valve (CV):

```
if |h| > Htol then
    if h < -Htol then      status = CLOSED
    if Q < -Qtol then      status = CLOSED
    else                   status = OPEN
else
    if Q < -Qtol then      status = CLOSED
else                       status = unchanged
```

where Htol = 0.0005 ft and Qtol = 0.001 cfs.

10. If the status check closes an open pump, pipe, or CV, its flow is set to 10^{-6} cfs. If a pump is re-opened, its flow is computed by applying the current head gain to its characteristic curve. If a pipe or CV is re-opened, its flow is determined by solving Eq. (D.1) for Q under the current headloss h, ignoring any minor losses.

11. Matrix coefficients for pressure breaker valves (PBVs) are set to the following: $p = 10^8$ and $y = 10^8 Hset$, where Hset is the pressure drop setting for the valve (in feet). Throttle control valves (TCVs) are treated as pipes with r as described in item 6 above and m taken as the converted value of the valve setting (see item 4 above).

12. Matrix coefficients for pressure reducing, pressure sustaining, and flow control valves (PRVs, PSVs, and FCVs) are computed after all other links have been analyzed. Status checks on PRVs and PSVs are made as described in item 7 above. These valves can either be completely open, completely closed, or active at their pressure or flow setting.

13. The logic used to test the status of a PRV is as follows:

```
If current status = ACTIVE then
    if Q < -Qtol                    then new status = CLOSED
    if Hi < Hset + Hml – Htol       then new status = OPEN
                                    else new status = ACTIVE

If curent status = OPEN then
    if Q < -Qtol                    then new status = CLOSED
    if Hi > Hset + Hml + Htol       then new status = ACTIVE
                                    else new status = OPEN

If current status = CLOSED then
    if Hi > Hj + Htol
    and Hi < Hset – Htol            then new status = OPEN
    if Hi > Hj + Htol
    and Hj < Hset - Htol            then new status = ACTIVE
                                    else new status = CLOSED
```

where Q is the current flow through the valve, Hi is its upstream head, Hj is its downstream head, Hset is its pressure setting converted to head, Hml is the minor loss when the valve is open (= mQ^2), and Htol and Qtol are the same values used for check valves in item 9 above. A similar set of tests is used for PSVs, except that when testing against Hset, the i and j subscripts are switched as are the > and < operators.

14. Flow through an active PRV is maintained to force continuity at its downstream node while flow through a PSV does the same at its upstream node. For an active PRV from node i to j:

$$p_{ij} = 0$$
$$F_j = F_j + 10^8 \text{Hset}$$
$$A_{jj} = A_{jj} + 10^8$$

This forces the head at the downstream node to be at the valve setting Hset. An equivalent assignment of coefficients is made for an active PSV except the subscript for F and A is the upstream node i. Coefficients for open/closed PRVs and PSVs are handled in the same way as for pipes.

15. For an active FCV from node i to j with flow setting Qset, Qset is added to the flow leaving node i and entering node j, and is subtracted from F_i and added to F_j. If the head at node i is less than that at node j, then the valve cannot deliver the flow and it is treated as an open pipe.

16. After initial convergence is achieved (flow convergence plus no change in status for PRVs and PSVs), another status check on pumps, CVs, FCVs, and links to tanks is made. Also, the status of links controlled by pressure switches (e.g., a pump controlled by the pressure at a junction node) is checked. If any status change occurs, the iterations must continue for at least two more iterations (i.e., a convergence check is skipped on the very next iteration). Otherwise, a final solution has been obtained.

17. For extended period simulation (EPS), the following procedure is implemented:

    a. After a solution is found for the current time period, the time step for the next solution is the minimum of:

    - the time until a new demand period begins,
    - the shortest time for a tank to fill or drain,
    - the shortest time until a tank level reaches a point that triggers a change in status for some link (e.g., opens or closes a pump) as stipulated in a simple control,
    - the next time until a simple timer control on a link kicks in,
    - the next time at which a rule-based control causes a status change somewhere in the network.

    In computing the times based on tank levels, the latter are assumed to change in a linear fashion based on the current flow solution. The activation time of rule-based controls is computed as follows:

- Starting at the current time, rules are evaluated at a rule time step. Its default value is 1/10 of the normal hydraulic time step (e.g., if hydraulics are updated every hour, then rules are evaluated every 6 minutes).

- Over this rule time step, clock time is updated, as are the water levels in storage tanks (based on the last set of pipe flows computed).

- If a rule's conditions are satisfied, then its actions are added to a list. If an action conflicts with one for the same link already on the list then the action from the rule with the higher priority stays on the list and the other is removed. If the priorities are the same then the original action stays on the list.

- After all rules are evaluated, if the list is not empty then the new actions are taken. If this causes the status of one or more links to change then a new hydraulic solution is computed and the process begins anew.

- If no status changes were called for, the action list is cleared and the next rule time step is taken unless the normal hydraulic time step has elapsed.

b. Time is advanced by the computed time step, new demands are found, tank levels are adjusted based on the current flow solution, and link control rules are checked to determine which links change status.

c. A new set of iterations with Eqs. (D.3) and (D.4) are begun at the current set of flows.

D.2 Water Quality

The governing equations for EPANET's water quality solver are based on the principles of conservation of mass coupled with reaction kinetics. The following phenomena are represented (Rossman et al., 1993; Rossman and Boulos, 1996):

Advective Transport in Pipes

A dissolved substance will travel down the length of a pipe with the same average velocity as the carrier fluid while at the same time reacting (either growing or decaying) at some given rate. Longitudinal dispersion is usually not an important transport mechanism under most operating conditions. This means there is no intermixing of mass between adjacent parcels of water traveling down a pipe. Advective transport within a pipe is represented with the following equation:

$$\frac{\partial C_i}{\partial t} = -u_i \frac{\partial C_i}{\partial x} + r(C_i) \qquad \text{D.5}$$

where C_i = concentration (mass/volume) in pipe i as a function of distance x and time t, u_i = flow velocity (length/time) in pipe i, and r = rate of reaction (mass/volume/time) as a function of concentration.

Mixing at Pipe Junctions

At junctions receiving inflow from two or more pipes, the mixing of fluid is taken to be complete and instantaneous. Thus the concentration of a substance in water leaving the junction is simply the flow-weighted sum of the concentrations from the inflowing pipes. For a specific node k one can write:

$$C_{i|x=0} = \frac{\sum_{j \in I_k} Q_j C_{j|x=L_j} + Q_{k,ext} C_{k,ext}}{\sum_{j \in I_k} Q_j + Q_{k,ext}} \qquad \text{D.6}$$

where i = link with flow leaving node k, I_k = set of links with flow into k, L_j = length of link j, Q_j = flow (volume/time) in link j, $Q_{k,ext}$ = external source flow entering the network at node k, and $C_{k,ext}$ = concentration of the external flow entering at node k. The notation $C_{i|x=0}$ represents the concentration at the start of link i, while $C_{i|x=L}$ is the concentration at the end of the link.

Mixing in Storage Facilities

It is convenient to assume that the contents of storage facilities (tanks and reservoirs) are completely mixed. This is a reasonable assumption for many tanks operating under fill-and-draw conditions providing that sufficient momentum flux is imparted to the inflow (Rossman and Grayman, 1999). Under completely mixed conditions the concentration throughout the tank is a blend of the current contents and that of any entering water. At the same time, the internal concentration could be changing due to reactions. The following equation expresses these phenomena:

$$\frac{\partial(V_s C_s)}{\partial t} = \sum_{i \in I_s} Q_i C_{i|x=L_i} - \sum_{j \in O_s} Q_j C_s + r(C_s) \qquad \text{D.7}$$

where V_s = volume in storage at time t, C_s = concentration within the storage facility, I_s = set of links providing flow into the facility, and O_s = set of links withdrawing flow from the facility.

Bulk Flow Reactions

While a substance moves down a pipe or resides in storage it can undergo reaction with constituents in the water column. The rate of reaction can generally be described as a power function of concentration:

$$r = kC^n$$

where k = a reaction constant and n = the reaction order. When a limiting concentration exists on the ultimate growth or loss of a substance then the rate expression becomes

$$R = K_b(C_L - C)C^{(n-1)} \qquad \text{for } n > 0, K_b > 0$$
$$R = K_b(C - C_L)C^{(n-1)} \qquad \text{for } n > 0, K_b < 0$$

where C_L = the limiting concentration.

Some examples of different reaction rate expressions are:

- *Simple First-Order Decay ($C_L = 0$, $K_b < 0$, $n = 1$)*

    $$R = K_b C$$

    The decay of many substances, such as chlorine, can be modeled adequately as a simple first-order reaction.

- *First-Order Saturation Growth ($C_L > 0$, $K_b > 0$, $n = 1$):*

    $$R = K_b(C_L - C)$$

    This model can be applied to the growth of disinfection by-products, such as trihalomethanes, where the ultimate formation of by-product (C_L) is limited by the amount of reactive precursor present.

- *Two-Component, Second Order Decay ($C_L \neq 0$, $K_b < 0$, $n = 2$):*

    $$R = K_b C(C - C_L)$$

    This model assumes that substance A reacts with substance B in some unknown ratio to produce a product P. The rate of disappearance of A is proportional to the product of A and B remaining. C_L can be either positive or negative, depending on whether either component A or B is in excess, respectively. Clark (1998) has had success in applying this model to chlorine decay data that did not conform to the simple first-order model.

- *Michaelis-Menton Decay Kinetics ($C_L > 0$, $K_b < 0$, $n < 0$):*

    $$R = \frac{K_b C}{C_L - C}$$

    As a special case, when a negative reaction order n is specified, EPANET will utilize the Michaelis-Menton rate equation, shown above for a decay reaction. (For growth reactions the denominator becomes $C_L + C$.) This rate equation is often used to describe enzyme-catalyzed reactions and microbial growth. It produces first-order behavior at low concentrations and zero-order behavior at higher concentrations. Note that for decay reactions, C_L must be set higher than the initial concentration present.

Koechling (1998) has applied Michaelis-Menton kinetics to model chlorine decay in a number of different waters and found that both K_b and C_L could be related to the water's organic content and its ultraviolet absorbance as follows:

$$K_b = -0.32 UVA^{1.365} \frac{(100 UVA)}{DOC}$$

$$C_L = 4.98 UVA - 1.91 DOC$$

where UVA = ultraviolet absorbance at 254 nm (1/cm) and DOC = dissolved organic carbon concentration (mg/L).

Note: These expressions apply only for values of K_b and C_L used with Michaelis-Menton kinetics.

- *Zero-Order growth ($C_L = 0$, $K_b = 1$, $n = 0$)*

    $R = 1.0$

    This special case can be used to model water age, where with each unit of time the "concentration" (i.e., age) increases by one unit.

The relationship between the bulk rate constant seen at one temperature (T1) to that at another temperature (T2) is often expressed using a van't Hoff - Arrehnius equation of the form:

$$K_{b2} = K_{b1} \theta^{T2-T1}$$

where θ is a constant. In one investigation for chlorine, θ was estimated to be 1.1 when T1 was 20 deg. C (Koechling, 1998).

Pipe Wall Reactions

While flowing through pipes, dissolved substances can be transported to the pipe wall and react with material such as corrosion products or biofilm that are on or close to the wall. The amount of wall area available for reaction and the rate of mass transfer between the bulk fluid and the wall will also influence the overall rate of this reaction. The surface area per unit volume, which for a pipe equals 2 divided by the radius, determines the former factor. The latter factor can be represented by a mass transfer coefficient whose value depends on the molecular diffusivity of the reactive species and on the Reynolds number of the flow (Rossman et. al, 1994). For first-order kinetics, the rate of a pipe wall reaction can be expressed as:

$$r = \frac{2 k_w k_f C}{R(k_w + k_f)}$$

where k_w = wall reaction rate constant (length/time), k_f = mass transfer coefficient (length/time), and R = pipe radius. For zero-order kinetics the reaction rate cannot be any higher than the rate of mass transfer, so

$$r = MIN(k_w, k_f C)(2/R)$$

where k_w now has units of mass/area/time.

Mass transfer coefficients are usually expressed in terms of a dimensionless Sherwood number (Sh):

$$k_f = Sh \frac{D}{d}$$

in which D = the molecular diffusivity of the species being transported (length^2/time) and d = pipe diameter. In fully developed laminar flow, the average Sherwood number along the length of a pipe can be expressed as

$$Sh = 3.65 + \frac{0.0668(d/L) Re\, Sc}{1 + 0.04[(d/L) Re\, Sc]^{2/3}}$$

in which Re = Reynolds number and Sc = Schmidt number (kinematic viscosity of water divided by the diffusivity of the chemical) (Edwards et.al, 1976). For turbulent flow the empirical correlation of Notter and Sleicher (1971) can be used:

$$Sh = 0.0149\, Re^{0.88} Sc^{1/3}$$

System of Equations

When applied to a network as a whole, Equations D.5-D.7 represent a coupled set of differential/algebraic equations with time-varying coefficients that must be solved for C_i in each pipe i and C_s in each storage facility s. This solution is subject to the following set of externally imposed conditions:

- initial conditions that specify C_i for all x in each pipe i and C_s in each storage facility s at time 0,

- boundary conditions that specify values for $C_{k,ext}$ and $Q_{k,ext}$ for all time t at each node k which has external mass inputs

- hydraulic conditions which specify the volume V_s in each storage facility s and the flow Q_i in each link i at all times t.

Lagrangian Transport Algorithm

EPANET's water quality simulator uses a Lagrangian time-based approach to track the fate of discrete parcels of water as they move along pipes and mix together at junctions between fixed-length time steps (Liou and Kroon, 1987). These water quality time steps are typically much shorter than the hydraulic time step (e.g., minutes rather than hours) to accommodate the short times of travel that can occur within pipes. As time progresses, the size of the most upstream segment in a pipe increases as water enters the pipe while an equal loss in size of the most downstream segment occurs as water leaves the link. The size of the segments in between these remains unchanged. (See Figure D.1).

The following steps occur at the end of each such time step:

1. The water quality in each segment is updated to reflect any reaction that may have occurred over the time step.

2. The water from the leading segments of pipes with flow into each junction is blended together to compute a new water quality value at the junction. The volume contributed from each segment equals the product of its pipe's flow rate and the time step. If this volume exceeds that of the segment then the segment is destroyed and the next one in line behind it begins to contribute its volume.

3. Contributions from outside sources are added to the quality values at the junctions. The quality in storage tanks is updated depending on the method used to model mixing in the tank (see below).

4. New segments are created in pipes with flow out of each junction, reservoir, and tank. The segment volume equals the product of the pipe flow and the time step. The segment's water quality equals the new quality value computed for the node.

To cut down on the number of segments, Step 4 is only carried out if the new node quality differs by a user-specified tolerance from that of the last segment in the outflow pipe. If the difference in quality is below the tolerance then the size of the current last segment in the outflow pipe is simply increased by the volume flowing into the pipe over the time step.

This process is then repeated for the next water-quality time step. At the start of the next hydraulic time step the order of segments in any links that experience a flow reversal is switched. Initially each pipe in the network consists of a single segment whose quality equals the initial quality assigned to the upstream node.

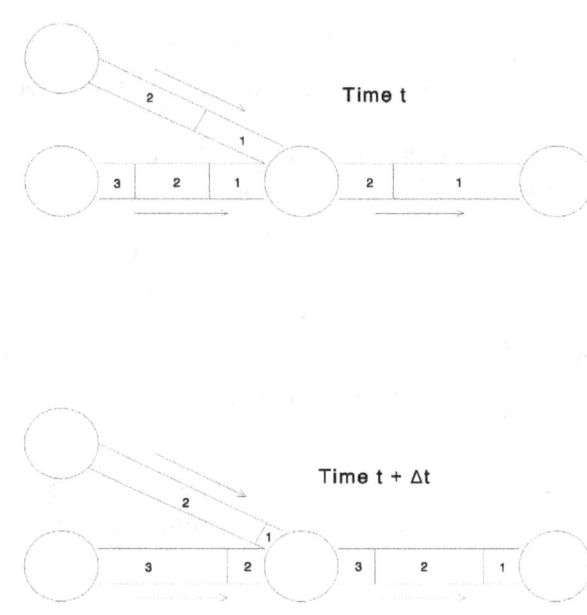

Figure D.1 Behavior of Segments in the Lagrangian Solution Method

D.3 References

Bhave, P.R. 1991. *Analysis of Flow in Water Distribution Networks*. Technomic Publishing. Lancaster, PA.

Clark, R.M. 1998. "Chlorine demand and Trihalomethane formation kinetics: a second-order model", *Jour. Env. Eng.*, Vol. 124, No. 1, pp. 16-24.

Dunlop, E.J. 1991. *WADI Users Manual*. Local Government Computer Services Board, Dublin, Ireland.

George, A. & Liu, J. W-H. 1981. *Computer Solution of Large Sparse Positive Definite Systems*. Prentice-Hall, Englewood Cliffs, NJ.

Hamam, Y.M, & Brameller, A. 1971. "Hybrid method for the solution of piping networks", *Proc. IEE*, Vol. 113, No. 11, pp. 1607-1612.

Koechling, M.T. 1998. *Assessment and Modeling of Chlorine Reactions with Natural Organic Matter: Impact of Source Water Quality and Reaction Conditions*, Ph.D. Thesis, Department of Civil and Environmental Engineering, University of Cincinnati, Cincinnati, Ohio.

Liou, C.P. and Kroon, J.R. 1987. "Modeling the propagation of waterborne substances in distribution networks", *J. AWWA*, 79(11), 54-58.

Notter, R.H. and Sleicher, C.A. 1971. "The eddy diffusivity in the turbulent boundary layer near a wall", *Chem. Eng. Sci.,* Vol. 26, pp. 161-171.

Osiadacz, A.J. 1987. *Simulation and Analysis of Gas Networks*. E. & F.N. Spon, London.

Rossman, L.A., Boulos, P.F., and Altman, T. (1993). "Discrete volume-element method for network water-quality models", *J. Water Resour. Plng. and Mgmt,,* Vol. 119, No. 5, 505-517.

Rossman, L.A., Clark, R.M., and Grayman, W.M. (1994). "Modeling chlorine residuals in drinking-water distribution systems", *Jour. Env. Eng.*, Vol. 120, No. 4, 803-820.

Rossman, L.A. and Boulos, P.F. (1996). "Numerical methods for modeling water quality in distribution systems: A comparison", *J. Water Resour. Plng. and Mgmt*, Vol. 122, No. 2, 137-146.

Rossman, L.A. and Grayman, W.M. 1999. "Scale-model studies of mixing in drinking water storage tanks", *Jour. Env. Eng.*, Vol. 125, No. 8, pp. 755-761.

Salgado, R., Todini, E., & O'Connell, P.E. 1988. "Extending the gradient method to include pressure regulating valves in pipe networks". *Proc. Inter. Symposium on Computer Modeling of Water Distribution Systems*, University of Kentucky, May 12-13.

Todini, E. & Pilati, S. 1987. "A gradient method for the analysis of pipe networks". *International Conference on Computer Applications for Water Supply and Distribution*, Leicester Polytechnic, UK, September 8-10.

www.ingramcontent.com/pod-product-compliance
Lightning Source LLC
Chambersburg PA
CBHW081145180526
45170CB00006B/1936